Functional Coatings for Corrosion Protection

Volume 2

Functional Coatings for Corrosion Protection

Volume 2

Dr. Vikas Mittal
Editor and Lead Author

Central West Publishing

This edition has been published by Central West Publishing, Australia

© 2021 Central West Publishing

All rights reserved. No part of this volume may be reproduced, copied, stored, or transmitted, in any form or by any means, electronic, photocopying, recording, or otherwise. Permission requests for reuse can be sent to editor@centralwestpublishing.com

For more information about the books published by Central West Publishing, please visit https://centralwestpublishing.com

Disclaimer
Every effort has been made by the publisher, editor and authors while preparing this book, however, no warranties are made regarding the accuracy and completeness of the content. The publisher, editor and authors disclaim without any limitation all warranties as well as any implied warranties about sales, along with fitness of the content for a particular purpose. Citation of any website and other information sources does not mean any endorsement from the publisher and authors. For ascertaining the suitability of the contents contained herein for a particular lab or commercial use, consultation with the subject expert is needed. In addition, while using the information and methods contained herein, the practitioners and researchers need to be mindful for their own safety, along with the safety of others, including the professional parties and premises for whom they have professional responsibility. To the fullest extent of law, the publisher, editor and authors are not liable in all circumstances (special, incidental, and consequential) for any injury and/or damage to persons and property, along with any potential loss of profit and other commercial damages due to the use of any methods, products, guidelines, procedures contained in the material herein.

A catalogue record for this book is available from the National Library of Australia

ISBN (print): 978-1-922617-06-4

Contents

Preface   XI

1. **Polymeric Pipeline Coatings for Oil and Gas Industry**   1

    1.1 Introduction   1
         1.1.1 Pipeline Coatings   1
         1.1.2 Requirements for Long Lasting Performance of Coatings   2
    1.2 Advancements in Polymeric Pipeline Coatings: A Brief Summary   2
    1.3 Polymeric Pipeline Coatings   4
         1.3.1 Polyolefin Based Pipeline Coatings   4
         1.3.2 Epoxy Based Pipeline Coatings   11
         1.3.3 Polyurethane Based Pipeline Coatings   18
         1.3.4 Fluoropolymer Based Pipeline Coatings   23
    1.4 Conclusions   28
    References   29

2. **Ionic Liquids and Poly(ionic liquid)s Based Coatings**   37

    2.1 Introduction   37
    2.2 Poly(ionic liquid)s Based Coatings   39
         2.2.1 Poly(ionic liquid)s Based Protective Coatings   39
         2.2.2 Poly(ionic liquid)s Based Sorbent Coatings   41
         2.2.3 Other Functional Coatings Based on PILs   43
    2.3 Ionic Liquids Based Coatings   45
         2.3.1 Ionic Liquids Based Coatings for Titanium   45
         2.3.2 Ionic Liquids Based Coatings for Copper   47
         2.3.3 Ionic Liquids Based Coatings for Steel   49

		2.3.4	Ionic Liquids Based Coatings for Aluminum	50
		2.3.5	Ionic Liquids Based Coatings for Magnesium Alloy	51
		2.3.6	Other Functional Coatings Based on ILs	53
	2.4	Ionic Liquids Based Electrodeposited Al Coatings		54
		2.4.1	Coatings for Steel	55
		2.4.2	Coatings for Copper	56
		2.4.3	Coatings for Gold	56
		2.4.4	Coatings for Magnesium Alloy	57
		2.4.5	Coatings for Neodymium Iron Boron (NdFeB) Magnet	57
		2.4.6	Coatings for Other Purposes	58
	2.5	Conclusions		60
	References			60
3.	**Carbon Based Coatings**			**73**
	3.1	Introduction		73
	3.2	Mesoporous Carbon Coatings		74
		3.2.1	Fabrication	74
		3.2.2	Characterization and Testing	77
		3.2.3	Applications	79
	3.3	Amorphous Carbon Based Coatings		80
		3.3.1	Fabrication	80
		3.3.2	Characterization and Testing	82
		3.3.3	Applications	85
	3.4	Coatings Based on Pyrolytic Carbon and Carbon Black		86
		3.4.1	Fabrication	86
		3.4.2	Characterization and Testing	88
		3.4.3	Applications	89
	3.5	Coatings Based on 'Carbon Alcohols' as Source		90
		3.5.1	Fabrication	90
		3.5.2	Characterization and Testing	91
		3.5.3	Applications	92
	3.6	Coatings Based on Graphene and Other Related Materials		92
		3.6.1	Fabrication	93

		3.6.2	Characterization and Testing	95
		3.6.3	Applications	98
	3.7	Miscellaneous Carbon Coatings		98
		3.7.1	Fabrication	99
		3.7.2	Characterization and Testing	100
		3.7.3	Applications	103
	3.8	Conclusions		103
	References			104
4.	**Self-Healing Anti-Corrosion Coatings Embedded with Graphene and Graphene Oxide based Containers Encapsulating Corrosion Inhibitor in Polyelectrolyte Complexes**			**109**
	4.1	Introduction		109
	4.2	Experimental		111
		4.2.1	Materials	111
		4.2.2	Preparation of Gr and GO Based Containers	111
		4.2.3	Zeta Potential	112
		4.2.4	Preparation of Composite Coatings	113
		4.2.5	Fourier-transform Infrared Spectroscopy (FTIR)	113
		4.2.6	Immersion Tests	113
		4.2.7	UV Weathering Analysis	113
		4.2.8	Electrochemical Analysis	114
	4.3	Results and Discussion		114
	4.4	Conclusion		130
	References			132
5.	**Self-healing Protective Coatings of Polyvinyl Butyral/Polypyrrole-Carbon Black Composite on Carbon Steel**			**137**
	5.1	Introduction		137
	5.2	Experimental		139
		5.2.1	Materials	139
		5.2.2	Preparation of PVB/PPyCB Formulations	139
		5.2.3	Application of Composite Formulations on Carbon Steel	139

		5.2.4	Electrochemical Measurements	140

	5.2.4	Electrochemical Measurements	140
	5.2.5	Immersion Test	141
	5.2.6	Characterization Techniques	141
5.3	Results and Discussion		142
	5.3.1	Morphology of the PVB/PPyCB Composite Coatings	142
	5.3.2	Electrochemical Measurements	144
	5.3.3	Immersion Test	155
	5.3.4	Protective Performance of Graphene Incorporated PVB/PPyCB Composite Coatings	158
5.4	Conclusions		162
References			164

6. UV Degradation of Polymer Coatings — 169

6.1	Introduction		169
6.2	Mechanism of UV Degradation		172
6.3	Different Polymer Coatings Systems		174
	6.3.1	PU Coatings	174
	6.3.2	Acrylic Coatings	175
	6.3.3	Epoxy Coatings	178
	6.3.4	Polyester Coatings	179
	6.3.5	Silicone Coatings	180
6.4	Summary and Outlook		180
References			180

7. Graphene for Corrosion Protection — 187

7.1	Introduction		187
7.2	Effect of Graphene on Electrochemical Properties of Carbon Steel in a Saline Media		189
	7.2.1	Materials and Methods	189
	7.2.2	Results and Discussion	191
7.3	Effect of Unmodified Graphene Platelets (UGP) on Electrochemical Properties of Polymer Coatings		195
	7.3.1	Materials and Methods	195
	7.3.2	Results and Discussion	198
7.4	Effect of Functionalized Graphene Platelets on Electrochemical Properties of Polymer		204

			Coatings	
		7.4.1	Materials and Methods	204
		7.4.2	Results and Discussion	205
	7.5	\multicolumn{2}{l}{Effect of Functionalized Graphene Platelets with Polyaniline on Electrochemical Properties of Polymer Coatings}	208
		7.5.1	Materials and Methods	208
		7.5.2	Results and Discussion	210
	7.6	\multicolumn{2}{l}{Conclusions}	213
	\multicolumn{3}{l}{References}	214
8.	\multicolumn{3}{l}{**Inhibition of Corrosion by Polyetherimide-Graphene Nanocomposite Coatings for Long Term Protection of Carbon Steel**}	**219**
	8.1	\multicolumn{2}{l}{Introduction}	219
	8.2	\multicolumn{2}{l}{Experimental}	222
		8.2.1	Materials	222
		8.2.2	Preparation of PEI-Graphene Coatings	222
		8.2.3	Characterization Techniques	223
		8.2.4	Immersion Tests	223
		8.2.5	Electrochemical Analysis	224
	8.3	\multicolumn{2}{l}{Results and Discussion}	224
	8.4	\multicolumn{2}{l}{Conclusions}	239
	\multicolumn{3}{l}{References}	240
9.	\multicolumn{3}{l}{**Effect of GO and rGO on the Corrosion Resistance of PMMA Nanocomposite Coatings**}	**245**
	9.1	\multicolumn{2}{l}{Introduction}	245
	9.2	\multicolumn{2}{l}{Experimental}	246
		9.2.1	Materials	246
		9.2.2	Substrate Preparation	247
		9.2.3	Preparation of Coatings	247
		9.2.4	Characterization Techniques	248
		9.2.5	Electrochemical Analysis	248
	9.3	\multicolumn{2}{l}{Results and Discussion}	249
	9.4	\multicolumn{2}{l}{Conclusions}	264
	\multicolumn{3}{l}{References}	264

10.	**Anti-microbial Polymeric Materials and Coatings**	**269**
	10.1 Introduction	269
	10.2 Different Anti-microbial Polymeric Materials and Coatings	270
	10.3 Conclusions	279
	References	280

Index **291**

Preface

Corrosion involves the degradation of metals due to the oxidation and reduction processes occurring during the interaction of the metallic surfaces with the aggressive environments. These electrochemical processes result in the impairment of materials' physical and mechanical properties such as strength and ductility. Application of coatings is the most widely used method for the corrosion protection of the metallic structures. Specifically, the polymeric coatings (or reinforced polymer coatings) protect the metal substrates from external corrosive agents by acting as an effective barrier. Besides that, depending on the morphology and structure of polymers, the coatings possess the properties like thermal, chemical and mechanical stability. These properties combined with the ability to strongly adhere to metal surfaces yield durable coatings with longer service lifetime compared with other metallic and inorganic coatings. The purpose of this book is to assimilate the recent advances in the field of functional coatings systems for achieving effective corrosion protection.

Chapter 1 reviews the progress in the polymers-based coatings used to protect the oil and gas pipelines. Chapter 2 summarizes the recent literature studies presenting various advancements in the development of different ionic liquids and poly(ionic liquid)s based coatings. Chapter 3 reviews the various types of coatings generated by employing the active carbon component in various structural forms. The most recent findings in the synthesis, characterization and application of these carbon-based coatings are highlighted. To explore the development of the nano-containers based on graphene, the layer by layer addition of the polyelectrolytes and corrosion inhibitor on the surface of graphene particles was explored in this study. The obtained nano-containers were subsequently embedded in the polymeric coatings to study the anti-corrosion behavior. In Chapter 5, the fabrication of stable self-healing organic coatings was reported by incorporating polypyrrole-carbon black composite pigment into polyvinyl butyral matrix. The photo-degradation of diverse polymer coatings (e.g., acrylic, epoxy, polyurethane, silicone and polyester coatings) has been discussed in detail in Chapter 6. Chapter 7 investigates the use of graphene as an anti-corrosion filler for polymeric coatings to replace chromates and other hazardous pigments. Chapter 8 presents a one-step procedure for the incorporation of graphene in polyetherimide coating formulations for effective corrosion protection of carbon steel substrates for long periods. In Chapter 9, a comparison of the role of graphene oxide and reduced graphene oxide in protecting the metal corrosion was carried out using poly(methyl methacrylate) as the coating matrix. Finally, Chapter 10 briefly overviews the anti-microbial polymer materials and coatings.

The support of all the chapter contributors is greatly acknowledged towards the successful accomplishment of the book. The book is dedicated to my family for their constant support.

Vikas MITTAL

1

Polymeric Pipeline Coatings for Oil and Gas Industry

1.1 Introduction

1.1.1 Pipeline Coatings

In oil and gas industry, about 10% of the overall production cost of oil and gas products is due to the pipelines and their maintenance [1]. In this respect, oil and gas industries continuously strive to diminish overall transmission pipelines cost. Pipeline coatings are immensely important for the maintenance of the pipelines and many research studies have focused on the development of functional coatings for the pipelines. Both external as well as internal pipeline coatings have been recommended to achieve protection of pipelines, thereby, ensuing a longer service life. More specifically, the internal protection is possible by the usage of liners, which can be used for applications demanding the corrosion protection as well as the rehabilitation of corroded surfaces. The usage of polymeric liners reduces the inner surface irregularities and roughness, thereby, creating substantial smoother surfaces. This lessens the maintenance cost of pipelines by offering improved flow efficiency of oil and gas through it [2,3]. In addition, the external protection of pipelines can be achieved through the usage of pipeline coverings, which provides sizeable anti-corrosive and wear resistance properties [4,5]. The selection of most effective coating is an important parameter that determines the durability of the lodged pipelines in the industries. In this respect, polymer based pipeline coatings have gained significant interest in oil, gas and petrochemical industries owing to their superior characteristics compared to other materials. These materials offer combination of scratch, wear and mechanical damage resistance along with good anti-corrosive properties to the pipelines, thus, creating enhanced life span. Among the various polymer categories useful for such coatings, epoxies, polyurethane (PU), polypropylene (PP), polyethylene (PE), copolymers of ethylene (e.g.: EBA and EVA), fluoropolymers, *etc.* have gained more

Anish M. Varghese and Vikas Mittal, Khalifa University of Science and Technology, Abu Dhabi, UAE
© *2021 Central West Publishing, Australia*

usage industrially due to their benefits in terms of both performance and cost [6].

1.1.2 Requirements for Long Lasting Performance of Coatings

The final performance of polymeric coatings depends greatly on the things happening during manufacture, application, transportation, installation and field operation stages of their lifetime. The freshly applied polymeric coatings successfully isolate the pipelines from the adverse environment conditions, thereby, preventing the corrosion [7]. Over the span of usage, the coating materials can undergo changes in the properties, which can affect the overall efficiency and performances of pipelines. The predominant reasons for these kinds of changes include permeation of air and water molecules in the coating, loss of adhesion and cohesion, disbondment with passage/prevention of cathodic protection current, increase in cathodic protection current, etc. [8]. Due to these reasons, many studies are currently being carried out to gain more insights about under coating corrosion in order to enhance the adhesion strength between the pipelines and polymeric materials, thereby, reducing the maintenance cost. The under coating corrosion is induced only when both water and oxygen are present together and lead to metallic dissolution/anodic effect, thus, resulting in corrosion. The important factors which decide the corrosion rate in pipelines include temperature, pressure, wind velocity profiles and composition of carrying fluids [9]. As the replacement of coatings for some parts of pipelines, for example for underground portions, is not possible in short duration of time, thus, oil and gas industries need long life of coating materials. Thus, due to these reasons, coating materials should possess properties such as resistance towards water molecules and moisture, pressure variations, bacteria and mushrooms, capillary effect of water, temperature variations, solvents (especially oils and their derivatives) and mechanical damages [10]. This review describes the progress in the common polymer based coatings used to protect oil and gas pipelines.

1.2 Advancements in Polymeric Pipeline Coatings: A Brief Summary

The first successful industrial pipeline coating was coal tar, which was introduced about 85 years ago. Up to 1970, coal tar was the major coating material and is still in use in some specific areas because of its outstanding moisture resistance. Subsequently, the use of coal tar enamel (CTE) was reported, which was initially a plasticized mixture of coal tar pitch, coal and distillates. To improve the properties of CTE, inert fillers were added to the composition [11]. Afterwards, four component CTE based coatings were developed in order to overcome the drawbacks of coal tar. Coal tar melts in the hot environment and becomes stiff in cold conditions and furthermore caused health problems. The four component CTE system consisted of primer, CTE, inner and outer wraps of glass fiber. In this system, epoxy was largely used as the primer, thus, it is also considered as one of the first polymer based pipeline coatings systems. Epoxy provided good adhesion, UV resistance and thermal stability to the system. Later, asphalt enamels were also introduced as a replacement for coal tar. These enamels provided good corrosion resistance, however, studies revealed the presence of lower amount of carcinogenic contents in asphalt enamels [12].

In the meantime, polymeric tapes also got industrial attention for use as pipeline coatings. The polymeric tape is a composite material of a soft elastomer based adhesive inner layer and an outer monolithic polymer layer. Commonly, the outer layer includes polyolefins, polyvinylchloride, butyl rubber, etc. as these polymers provide good mechanical strength along with outstanding thermal, electrical and corrosion resistance to the coatings [13]. Following this, heat shrinkable tape coatings were introduced in 1980's. Generally, these consist of a high shear strength hot melt adhesive of thermoplastic with a thick radiation crosslinked polyethylene based backing. These coatings need preheated pipeline surfaces to melt the adhesive layer and, thus, generate strong adhesion with the pipeline surfaces. Heat shrinkable structures were obtained through the heating of crosslinked polyethylene backing with the aid of propane torches from outside [14,15].

In the early 1960s, fusion bonded epoxy (FBE) was generated as an effective pipeline protective coating. Extensive developments concerning the performance improvements of FBE have been carried out since then. Generally, it is a 100% solid thermosetting epoxy powder, which offers strong adhesion to metallic surface through heat induced crosslinking [16,17]. Also, spray-applied liquid coatings have received attention as effective pipeline coating systems. Normally, these coatings are based on high build epoxy, polyurethane or a combination of two along with curing agents [18].

Following the developments of both epoxy and polyolefin based coatings, coating systems based on the combination of these two also generated industrial interest. Both two-layer and three-layer PE and PP coatings were developed. These coatings presented comprehensive characteristics of both epoxies and polyolefins [19]. Following the modifications of PE based coatings, high performance composite coatings (HPCC) were developed and applied for pipeline protection. It is a three component powder system consists of FBE primer, chemically modified polyethylene tie layer and polyethylene outer layer [13]. Biopolymer based coatings have also been explored for the purpose of generating anti-corrosion property. Figure 1.1 shows the electrochemical impedance spectroscopy (EIS) plots of polyvinylbutyral (PVB) and chitosan (Ch) coatings cross-linked with glutaraldehyde (Glu), developed for potential application as external pipeline coatings [20]. More recently, epoxy nanocomposite coatings, PU nanocomposite coatings and hybrid fluoropolymer resins have been introduced to enhance the performances of epoxy, PU and fluoropolymer based pipeline coating systems.

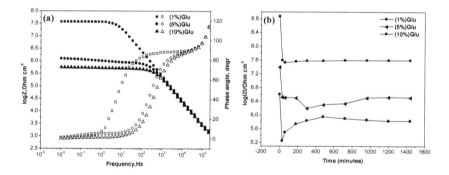

Figure 1.1 EIS plots of PVB_Ch/x%Glu_PVB coating having different percentages of glutaraldehyde; (a) Bode and phase plots obtained after 2 h immersion in 0.3 M salt solution and (b) logZ at low frequency from the Bode plot vs. time of immersion. Reproduced from Reference 20 with permission from Springer.

1.3 Polymeric Pipeline Coatings

1.3.1 Polyolefin Based Pipeline Coatings

Polyolefin is a group of thermoplastic polymeric materials obtained from the polymerization of monomer olefin with a general formula of C_nH_{2n}. These classes of polyolefin based coatings include either PE or PP. These polymers possess excellent corrosion resistance and outstanding mechanical strength for pipeline protection. Furthermore, health and safety issues associated with these polymers are minimal as compared to other materials. However, their low degree of adhesion or bonding to metallic surfaces is a significant limitation [21,22]. In order to overcome this, significant developments have been achieved in the last decades in this area. Among the various advancements, two-layer and three-layer polyolefin based coatings gained more industrial interest.

Two-layer polyolefin coatings were introduced in 1960's. This coating system provides a combination of bonding and corrosion protection. This consists of an extruded polyolefin over coating and an inner layer of adhesive or sealants. The application of polyolefin coatings need the adhesive layer based on butyl or asphalt pipeline. Adequate flow of adhesive inner layer over pipelines upon heating is necessary to get uniform coating and which is then covered with polyolefin extrudates (side extruded or cross head extruded polyolefin) based on either high density PE (HDPE) or PP [15,23]. Three-layer polyolefin coatings are one of the most acceptable multilayer polymeric coating systems and have been in application since 1980's. These coatings are composed of an inner layer of either FBE or liquid epoxy, a tie layer of olefin copolymer adhesive and an outer layer of polyolefin. The inner layer of primer is applied with the aid of electrostatic spraying, followed by heating of pipeline surfaces to suitable temperature to obtain the uniform coating. Afterwards, the tie layer of adhesive is applied over primer surface using either side extrusion or spraying, followed by final coating of polyolefin outer layer [15,23].

PE Based Pipeline Coatings

PE is one of the most commonly used commodity thermoplastic polymeric material with a chemical formula of $(C_2H_4)_n$. Among the various grades of PE, high density PE (HDPE) and medium density PE (MDPE) have received more acceptance for pipeline coating applications. Both of these grades are prepared using Ziegler-Natta catalysts. PE based pipeline coatings provide outstanding performance such as good mechanical properties, superlative corrosion resistance, fine chemical stability and inexpensiveness. However, PE possesses low softening point, which usually reduces the applications overreaching the temperature of 80°C [24].

Due to its excellent properties, PE has been widely used in many kinds of pipeline coatings formulations and has undergone diverse modifications in order to suit the pipeline coating applications. The history of PE as a successful pipeline coating material has started through its usage as polymeric tapes. As mentioned earlier, polymeric tape is a coating material composed of an adhesive inner layer and a monolithic polymer outer layer. In PE tapes, outer covering layer of PE offers good mechanical properties as well as combination of corrosion, electrical and chemical resistance properties to the coatings. Normally, a soft elastomeric adhesive inner layer is used to enhance the adhesion of PE outer layer over pipelines, which serves as primer. Specifically, the anti-corrosive PE outer layer is wrapped throughout the adhesive coated pipelines [5]. As a continuation of PE tapes, PE based heat shrinkable tape coatings were developed. As also mentioned earlier, it has an inner lining of thermoplastic hot melt adhesive and a monolithic radiation cross-linked PE cover coating. In this system, the outer coverings are applied over the pipelines with hot melted adhesive inner layer. Heat shrinkable structure can be generated through the heating of radiation crosslinked PE from the outside with the aid of propane torches. Moreover, a three component system of PE has also been developed, which uses an additional epoxy primer [11].

As a result of continuous developments in PE adhesion over the pipeline surfaces, two-layer and three-layer PE coatings were introduced. More details of these are available in the earlier section of this review. Among these two coating systems, three-layer PE coatings have received more interest. Structurally, these are composed of epoxy inner lining, PE copolymer adhesive tie layer and PE surface layer. The performance of such a system results from the combined contribution from individual layers. Epoxy lining provides cathodic disbonding resistance and improved cohesion to the system. Due to good moisture, oxygen and chemical resistance as well as excellent mechanical properties of PE, it acts as an outstanding protective barrier. Generally, epoxy primers are applied by electrostatic spraying whereas middle and PE surface coatings are employed on the pipelines using extrusion [25,26].

Kamimura *et al.* [27] studied the cathodic disbonding mechanism of three layer PE pipe coating consisting of liquid epoxy primer, maleic acid anhydride modified PE adhesive layer and PE protective layer. The impact of coating thickness, dissolved oxygen, sodium chloride (NaCl) concentration and cathodic potential on the cathodic disbonding of PE in NaCl solution under an elevated temperature of 65 °C was investigated. Moreover, the migration routes of cations, water and oxygen molecules to the coatings were studied. From the results of the disbonding test over 14 days, it was observed that the disbonding radius (r) of coatings decreased with increase in PE thickness from 0.9 to 8.5 mm. The migration of oxygen and water molecules through PE contributed to disbanding, however, the increase in thickness of PE was observed to shield the migration of oxygen and water molecules through it and thereby resulted in a decrease in r value. The excellent chemical resistance of PE prevented the sodium migration for all coating thicknesses. At the same time, the migration of sodium through the interface of coatings and steel was observed. Furthermore, the authors reported strong disbonding inhibiting effect of coatings at elevated temperature with strong evolution of hydrogen along with a cathodic potential of -1500 mV$_{SCE}$ [27]. In another study, Guermazi et al. [28] reported the hygrothermal ageing consequences of HDPE pipeline coatings on physico-chemical, mechanical and tribological properties. The authors submerged the HDPE pipeline coatings in distilled water as well as synthetic sea water (saline solution) at an elevated temperature of 70 °C to accelerate the ageing process. The existence of a low diffusion process in both the solvents was observed and a comparatively higher diffusion of water molecules was found in distilled water. A decrease in glass transition temperature (T_g) of HDPE coatings was observed with increase in immersion duration, which resulted from the increase in flexibility of amorphous molecular part of the coatings through the plasticizing effect of water molecules. The decreased tensile strength and elastic modulus values of coatings after hygrothermal ageing revealed the deterioration of mechanical properties. Furthermore, the authors reported that decrease in wear resistance for aged coating samples, which was attributed to the plasticizing effect of the penetrated solvent molecules in the amorphous part [28]. In a similar study, Guermazi et al. [6] investigated the effect of hygrothermal ageing time and temperature on the structural as well as mechanical properties of HDPE pipeline coatings. The samples were immersed in synthetic sea water at varying temperatures of 23, 70 and 90 °C for different immersion periods up to several months. An increase in solvent diffusion in to the coatings was observed with increase in temperature. Also, deterioration in mechanical properties such as tensile modulus, tensile strength and stress at 500% of strain was observed with enhancement in ageing parameters such as temperature and time. Furthermore, the authors displayed that the degradation behavior of aged samples through the structural variations using Fourier transform infrared (FTIR) spectroscopy. FTIR spectroscopic analysis demonstrated a decrease in peak intensities as well as vanishing of some functional groups in the coating materials after hygrothermal ageing [6].

Tribological behavior of pipeline coatings is an important parameter that determines the application in oil and gas industries. Guermazi *et al.* [29] investigated the friction and wear responses of both unaged and hygrothermally aged HDPE coating samples under varying applied loads and testing times at room temperature in a pin-on-disk tribometer. The applied load and test duration had strong impact on the friction coefficients and wear resistance of coatings. Increasing tendency of friction coefficients and wear volume were observed with increase in load. Also, the authors reported the independency of friction coefficient and the dependency of wear volume on the accelerated ageing. The remarkable deterioration of wear resistance was noticed for aged samples, particularly for samples aged for long durations. Moreover, the authors proposed a wear mechanism of coatings and confirmed the existence of direct relationship between wear volume and energy dissipation using an energetic quantitative approach [29]. In another study, the response of PE coatings towards the scratch damage was investigated [30]. For this purpose, unaged and aged coating samples were analyzed under varying scratch parameters of sliding velocity, angle of attack and applied normal load at room temperature. A decrease in friction coefficient with increasing sliding velocity and an increase in friction coefficient with increasing attack angle of the indenter were reported. The variation of applied load had no significant effect on the friction coefficient. The existence of combined permanent viscous as well as temporary elastic deformations during the application of scratch load was reported because of the viscoelastic behavior of PE. Furthermore, variation in wear volume of the coatings with increase in attack angle of indenter and applied normal load was noticed. The degradation behavior of samples was examined using hygrothermally aged coatings and remarkable decline in the wear resistance of samples was observed [30].

In another study, Samimi *et al.* [9] studied the causes of corrosion in three layer PE coated steel pipelines. The authors confirmed the influence of initial adhesiveness and contact environment on the performance of coatings. Moreover, the authors demonstrated the impact of moisture as well as cathodic disbonding resistance of coatings on the durability [9]. In another study, the occurrence of disbonding at the steel/FBE interface of the three-layer PE pipeline coating was investigated [31]. The authors reported the existence of under cured FBE layer when the applied temperature for adhesive-FBE bonding was low. In addition, a weak bonding between the FBE and adhesive was also observed when the temperature was high because of the fully crosslinked structure of FBE.

As mentioned earlier, high performance composite coating (HPCC) is a multi-component powder system composed of FBE primer, tie layer of chemically modified PE copolymer adhesive and outer layer of MDPE. In order to reduce the interlayer delamination and, thus, to create a single coating structure, a blend of adhesive and FBE is used as the tie layer for this system. As a result, a strong interlocked structure without well-defined interfaces is formed between the components. This results from the structural similarities of adhesive layer and PE and which helps to intermingle effectively with the structures containing FBE [13,32]. Howell *et al.* [33] evaluated physical, chemical and mechanical properties of coating membranes and coated steel pipes based on HPCC. The authors analyzed microstructure, impact strength, adhesion, water permeability, resistance towards cathodic disbondment and electrochemical impedance of coating samples. The presence of PE in the HPPC coatings was observed to result in a uniform coating structure with improved water and chemical diffusion resistance. Further, superior adhesion of HPCC to the steel pipelines was observed according to both ASTM and CSA standards. The authors also observed good impact energy of 10.2 J at low temperature conditions (0 °C). The results of impedance measurements underlined outstanding anti-corrosive characteristics of HPCC because of capillary behavior. Moreover, a little cathodic disbondment was reported for HPCC. In another study, Singh *et al.* [34] compared the properties of powder coated HPCC and conventional 3 layer PE coatings. Application of HPCC was beneficial for special needs such as coatings for raised welds and thickness adjustments. Excellent adhesion as well as good barrier properties for HPCC were observed. The authors also reported excellent corrosion resistant behavior and mechanical properties for the coating. In another work, Guan *et al.* [32] also observed outstanding cost effective performance of HPCC as compared with typical coating systems. The performance was confirmed through net lifelong performance-cost satisfaction investigations of miscellaneous coating systems [32].

In order to resolve the usual complications associated with the conventional three layer PE coatings such as weld tenting and low coating thickness on the outside weld parts, Lam *et al.* [35] developed a new system of polyethylene coatings. The authors introduced a side extruding HDPE above the graded structure PE coating (GSPE), which led to uniform coating over the external weld, thus, avoiding weld tenting. Moreover, the authors observed considerably low residual stresses on the coatings, which resulted in low temperature flexibility and good impact strength and, thus, improved adhesion of these coatings [35].

PP Based Pipeline Coatings

PP is a versatile semi-crystalline commodity thermoplastic polymer with a chemical formula of $(C_3H_6)_n$, prepared using Ziegler-Natta catalysts [36]. PP based pipeline coatings have been in use since 1980's and have gained attention because of their enhanced mechanical and thermal properties along with advantageous chemical resistance as compared to conventional PE and epoxy based systems [37]. Till now, PP has been used in diverse pipeline coatings include polymeric tapes, heat-shrinkable sleeves and two-layer and three-layer PP coatings.

Suzuki *et al.* [38] reported PP coated steel pipes suitable for oil and gas transmission at -30°C to 120°C temperature range. Structurally, the system contained steel pipe, modified polyolefin based adhesive inner layer and PP outer layer. The authors developed the PP based pipeline coatings to resolve the problems of PE based system and to obtain good coating performance even at higher temperatures. PP coatings exhibited good chemical stability and elevated softening temperature, however, inferior mechanical properties were observed at lower temperatures when compared to PE coatings. As a continuation of advancements in PP based pipeline coatings, Guidetti *et al.* [39] made use of three-layer PP coating system for oil and gas transmission pipelines. The three-layer coating system comprised of epoxy resin inner layer, modified PP copolymer based tie layer and outer covering of PP. Significant bonding of epoxy resin with oxides of metal surfaces as well as polar groups of adhesive tie layer was observed. Also, the existence of compatible structures was observed due to the structural resemblance of tie layer and PP outer layer. Thus, appreciable performance of the three-layer PP coatings was obtained due to the synergistic effect of epoxy resin and PP. Epoxy resin contributed adhesion, excellent cathodic disbonding resistance and good interfacial properties to the system, whereas PP developed outstanding physical and mechanical properties, high temperature stability and resistance towards chemicals, corrosion and water permeation. In another study, Moosavi *et al.* [40] examined the feasibility of three-layer PP pipeline coatings at elevated temperatures (more than 100°C) for long term performance. Total disbondment of coatings from the steel pipes as well as the serious cracking and tearing were observed. This was attributed to the low temperature resistance of FBE layer in the three-layer PP pipeline coatings. Further, the increased operating temperatures and heat applied maintenance activities were observed to reduce the life span of the coating. The authors observed decrease in mechanical properties on changing the FBE layer material to harder one. The authors recommended the use of FBE primer having a glass transition temperature (T_g) in the range of pipeline design temperature [40]. In another work, Moosavi *et al.* [41] investigated the failure history of the untimely failed three-layer PP gas pipeline coatings at high temperature. The authors observed the cracking of polypropylene top layer and the disbondment of the coatings from the surface of steel pipe. The cracking of PP outer layer was observed to result from the net effect of thermo-oxidative degradation and high residual stresses, whereas the disbondment of three-layer PP coating resulted from the adhesion loss as well as high residual stresses [41]. Overall, PP has emerged as a useful material for the pipeline coatings.

1.3.2 Epoxy Based Pipeline Coatings

Epoxies are high performance thermosetting polymeric material with wide range of applications especially for polymer based coatings. These materials are prepared by the condensation reaction of bisphenol-A (BPA) and epichlorohydrin (ECH), which creates molecules having two or more epoxide groups [42]. As compared with other polymeric materials, epoxy based coating materials provide excellent performance profile, which makes these coatings the materials of choice in many applications. Epoxy based coatings impart good adhesion, chemical resistance, outstanding cathodic protection, stress cracking resistance and excellent resistance towards microorganisms [16]. Thus, epoxy based coating systems have been developed in various forms and applications. These materials find use in applications such as liquid epoxy coatings, fusion bonded epoxy (FBE) coatings, base primer coatings for many conventional coating systems and so on. Epoxy based coatings are well accepted for pipeline protection through both external and internal coating applications. The excellent properties such as suitable coating thickness in a single step, 100% solid nature, corrosion resistance, easy pipeline cleaning and improved fluid flow efficiencies of epoxy based coatings make them advantageous for inner liners [3,43].

Coatings based on liquid epoxies comprise of two-part system of epoxy resin and curing agent, where the curing agent normally used is either polyamine or polyamide. Also, liquid epoxies can be used in combination with polyurethane as a spray applied coating system [18]. Epoxies also find applications as base coat primer for many coating systems including coal tar enamels, two-layer and three-layer polyolefin coating systems, HPCC and so on because of their excellent adhesion characteristics to steel surfaces. Another significant application area of epoxies is FBE, available as single layer and dual coat FBE. Generally, FBE is a heat curable one part powder form of thermosetting epoxy resin and exhibits good adhesion, fine surface finish and outstanding resistance towards abrasion, chemicals and soil stress. Single layer FBE coatings have been used since 1960's and are a single layer monolithic structure of FBE. These are applied over the well cleaned and preheated pipelines using electrostatic spraying. Dual coat FBE coatings are advanced form of single layer FBE coatings with versatile performance profile and have been in use since 1990's. Structurally, dual coat FBE comprise of top and base FBE coatings. This multilayer structure of coating system imparts many advantages to the pipelines such as resistance to mechanical damages caused by impact load, abrasion, gouge, friction with organic and inorganic surfaces, etc., along with ultraviolet or weathering resistance as well as high temperature performance [43].

The friction factor of the flowing gas through the pipelines greatly depends on the nature of internal coating. Yang *et al.* [3] conducted an aerodynamic evaluation of the internal epoxy coatings of natural gas pipelines to study this phenomenon. The authors compared the Colebrook-White equation based numerical friction factors with the data obtained from both field tests and model experiments. The authors observed consistency in both numerical and experimental readings, which proved the usefulness of numerical model for the analysis of epoxy internal coated gas pipelines. The application of internal epoxy coatings for the gas pipelines was observed to lower the frictional pressure drop, thereby, reducing the payback time [3]. Wei *et al.* [44] reported the influence of flow conditions on the performance degradation of FBE coatings in corrosive environments. For this purpose, three miscellaneous FBE coatings in 3% NaCl solution at 60 °C were studied under flowing and steady conditions and the degradation behavior was analyzed using electrochemical impedance spectroscopy (EIS). The authors observed strong effect of subjective conditions and flowing environment on the performance deterioration of protective coatings. At flowing conditions, the ions present in the corrosive environments diffused easily through the coatings when compared to water molecules.

In many recent studies, different advanced epoxy based coatings systems have also been reported, which confirm high potential of these systems for use in pipeline coatings [45,46]. For instance, Luo and Mather [45] reported an advanced system based on shape memory assisted self-healing coatings. This was achieved by distribution of electrospun thermoplastic polycaprolactone in a shape memory epoxy matrix. As shown in Figure 1.2, the damage to the coatings could be healed automatically on the application of heat. In another study, Augustyniak *et al.* [46] reported smart epoxy coatings here early detection of steel corrosion could be gauged through fluorescence. It was achieved through fluorescence indicator added in the epoxy coatings systems, which formed complex with the ferric ions generated during corrosion process. The indicator was observed to become fluorescent in and around the areas where the corrosion was initiated, however, this was achieved before any observable damage to the metal occurred, thus, indicating the advanced nature of detection. Figure 1.3 also shows the performance of the indicator in various test conditions. The fluorescence effect was clearly visible in the images, which resulted in early detection of any corrosion process using the advanced coating system. In another study, Wei *et al.* [47] reported the effect of carbon black (CB) on the performance improvement of FBE coatings under corrosive environments. The FBE coatings with varying CB loadings (0.5-4 wt %) were immersed in a 3% NaCl solution at room temperature. The degradation behavior of CB filled FBE pipeline coatings was analyzed using EIS, thermo-gravimetry, differential scanning calorimetry and visual inspections. FBE coating with CB concentration above the percolation value exhibited peculiar electrochemical behavior. T_g value was observed to increase significantly after crossing the percolation value. Further, the addition of CB in to FBE improved its barrier properties through the CB network formation, which correspondingly contributed to the enhancement of the coating's performance.

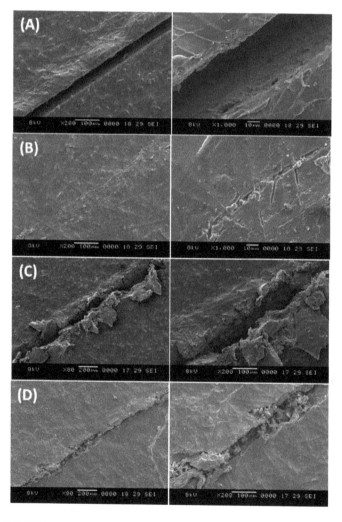

Figure 1.2 SEM micrographs of (A) cracked coating, (B) crack coating after self-healing, (C) scribed coating, and (D) scribed coating after self-healing. Reproduced from Reference 45 with permission from American Chemical Society.

Goertzen *et al.* [48] investigated the creep behavior of the coating system based on carbon fiber-epoxy composite to understand the long-term deflection and failure behavior. Creep tests were carried out at both room temperature and elevated temperatures with the help of respective tensile and flexural creep analysis. The authors observed consistent creep curves in both experiments. Tensile creep tests revealed outstanding creep rupture resistance of composite coatings and the coating material had the capability to withstand a load of 77% of the ultimate tensile strength (UTS) for up to 1600 h. Elevated temperature flexural creep analysis was performed on a dynamic mechanical analyzer (DMA) with a temperature profile of 30 to 75 °C. The authors predicted the creep levels of coatings extrapolated to the 50 years life span and observed the stress value of 84% of UTS at 30 °C and 42% of UTS at 50 °C. Further, the modulus reduction of 18% at 30 °C and 58% at 50 °C after 50 years of composite coating life was predicted. Another study by Alamilla *et al.* [49] reported the failure inspection as well as the mechanical behavior of FBE coated oil pipelines. The authors followed five different methodologies to examine the failure behavior including visual inspection, environmental analysis, interfacial characterization, mechanical and metallurgical analysis. It was revealed that the existence of both iron oxide and iron sulfide accelerated corrosion mechanisms on the FBE coated oil pipelines. Also, the authors underlined that the leakage caused ductile type failure on the oil pipelines, which was in good agreement with the previous reported works.

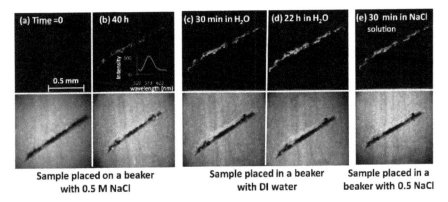

Figure 1.3 Scribed area on the coated sample as a function of different exposure time to different corrosive environments. Top row: fluorescent images and bottom row: digital camera images. Reproduced from Reference 46 with permission from American Chemical Society.

Zhou et al. [50] developed high T_g FBE coatings to protect the oil/gas pipelines at inflated service temperatures. The higher T_g of final coatings was obtained through the incorporation of brominated epoxy resin in the composition of thermoset epoxy powder coatings. In the coating system, the T_g enhanced up to 159 °C, which was about 39% greater than that of general FBE. Enhanced cathodic disbondment resistance and water soak adhesion were observed when the coatings were analyzed for 28 days at 95 °C. The system exhibited enhanced UV resistance, flexibility and better impact properties as compared to general FBE. In another study, Moon et al. [51] also compared the performance of high T_g FBE and general FBE. The experimental results confirmed the outstanding performance of high T_g FBE at elevated temperatures. To protect and strengthen the pipelines, Duell et al. [52] also introduced a fiber reinforced polymer (FRP) system based on carbon fiber and epoxies. The authors used an epoxy putty to eliminate the surface roughness of defected surfaces and carbon fiber-epoxy composites as the protecting covering. Various defect geometries on the pipeline surfaces were detected using finite element analysis. Both modeling and field tests were applied to confirm the effectiveness of carbon fiber-epoxy composite based coatings. Furthermore, a little effect of varying defect geometries on the failure pressure was observed. In another study, Yuan et al. [53] reported self-healing polymer coatings based on epoxy/mercaptan as healant. In this system, epoxy and mercaptan hardener were encapsulated individually and were subsequently embedded in epoxy. The material exhibited significantly enhanced self-healing performance (Figure 1.4) even at much lower capsule content. For instance, 43.5% healing efficiency was observed with 1 wt % capsules and 104.5% healing efficiency was reported with 5 wt % capsules. The healing occurred at or below room temperature, thus, further confirming the potential of the developed system to be effective for pipeline protection. Much better balance between strength and toughness of the healed system could be achieved.

In another study, Bakhshandeh et al. [54] reported the development of anti-corrosive organic-inorganic hybrid coatings based on epoxy-silica nanocomposites. The coatings were fabricated using silane functionalized diglycidyl ether of bisphenol A epoxy resin and pre-hydrolyzed tetraethoxysilane (TEOS) with the aid of 3-aminopropyl triethoxysilane (APTES) as coupling agent. The extent of compatibility between organic and inorganic phases was varied using different amounts of APTES. It was noticed that the smaller silica domains were generated for the coating system with epoxide to amine ratio of 4:1. Hybrid composites with 12.5 wt% of TEOS content exhibited optimal adhesion strength as well as micro-hardness. The enhanced barrier performance and the resultant

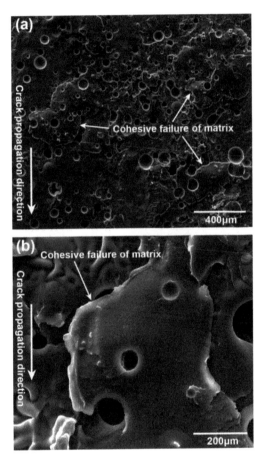

Figure 1.4 SEM images of the fracture surface of a healed specimen, containing 10 wt % epoxy-loaded capsules and 10 wt % hardener-loaded capsules. Reproduced from Reference 53 with permission from American Chemical Society.

corrosion resistance of coatings were attributed to the creation of silica intermediate layer at the coating-substrate interface. In another study reporting the advancement of epoxy based coating systems, Weng et al. [55] generated advanced anti-corrosion coatings by mimicking fresh plant leaves, thorough the combination of superhydrophobicity and redox catalytic capability. The authors generated superhydrophobic elecroactive epoxy (SEE) coating on steel using nano-casting method from the surface structure of Xanthosoma sagittifolium leaves. Due to such structured coating, the anti-corrosion performance was observed to enhance significantly as compared to the smooth coating on the steel substrate. Figure 1.5 exhibits the atomic force microscopy (AFM) images of the SEE surface topography.

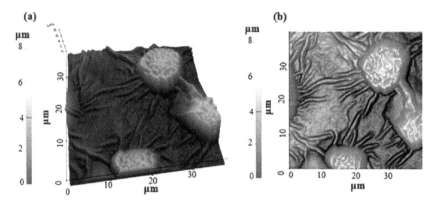

Figure 1.5 AFM images of the SEE topography; (a) 2-dimensional and (b) 3-dimensional. Reproduced from Reference 55 with permission from American Chemical Society.

1.3.3 Polyurethane Based Pipeline Coatings

Polyurethanes (PUs) are class of polymeric materials first introduced in 1937 and are available in thermoplastic, thermoset and elastomeric forms. These materials find wide variety of commercial and technical applications because of their attractive property profile. Commonly, these materials are prepared by the polyaddition reaction between polyisocyanates and macro-polyols. The final structure of PU is driven by the nature of isocyanates and polyols, along with the extent of crosslinking [56-58]. PU pipeline coatings offer enhanced corrosion, abrasion, scratch, environmental degradation and tear propagation resistances along with better adhesion, low temperature impact strength, adherence to health and safety stipulations and rapid curing speed. Due to the crosslinked structure in the thermoset PU coatings, further improvement in tensile strength, abrasion resistance and chemical stability of the coatings is achieved [22]. Also, rigid PU foams are widely used for the thermal insulation of pipelines in the oil and gas industries owing to their enhanced insulating properties.

Ghosal *et al.* [59] reported soya polyurethane based silica hybrid composite coatings for anti-corrosion performance. In-situ generation of silica in the polymer led to enhanced thermal, physico-mechanical and corrosion resistance of the hybrid coatings. Figure 1.6 shows the potentiodynamic polarization (PDP) studies of the composites in comparison with pure polymer and pure substrate. The corrosion protection ability of the coatings enhanced with increasing the amount of silica in the composites. The enhanced corrosion protection efficiency of the composite coatings was attributed to the blocking effect and good adhesion of coating with the substrate surface. In another study, Chattopadhyay *et al.* [60] studied

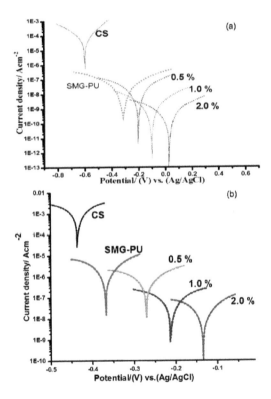

Figure 1.6 PDP curves of (a) CS, (b) SMG-PU, and composites with (c) 0.5%, (d) 1% and (e) 2% silica content in 3.5 wt % NaCl medium (SMG: soy oil monoglyceride, CS: carbon steel). Reproduced from Reference 59 with permission from American Chemical Society.

the effect of chain extender on the phase mixing as well as coating performance of polyurethane ureas. It was concluded that the bulky diol chain extenders helped to enhance phase mixing. Sufone (SUL) based chain extender was also observed to enhance mechanical properties of the polymer, along with surface segregation. Figure 1.7 also demonstrates the G' plots for SUL as a function of temperature at various angular frequencies. Another study by Samimi et al. [61] reported the properties and the application of 100% solid PU based pipeline coatings. The authors reported outstanding performance such as good adhesion, excellent resistance to corrosion, stroke, chemicals and frication, satisfactory flexibility and high temperature resistance. Furthermore, non-toxicity along with easy handling and fast curing of the coatings was also reported. Guan et al. [62] also investigated the performance of 100% solid rigid PU coatings used for the welded pipeline joints and rehabilitation of oil and gas pipelines. These advanced coatings were of castable or sprayable type and exhibited much faster curing. The authors also confirmed the superior performance of the coatings in comparison with other commercially available coating systems based on PU and other matrices. Furthermore, the coatings exhibited adherence to health and safety regulations as well as easy handling due to absence of volatile organic compounds (VOC), cold temperature curing, low curing time and balanced viscosity.

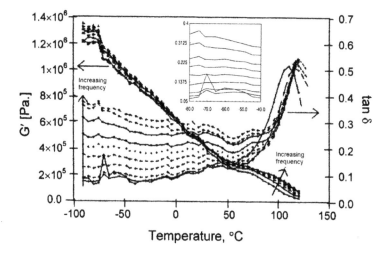

Figure 1.7 Variation of G' for SUL as a function of temperature at various angular frequencies (1, 5, 10, 20, 30, 40, and 80 rad/s). Reproduced from Reference 60 with permission from American Chemical Society.

Kong et al. [63] developed high-solid PU coating system using vegetable oil derived polyols. The generation of polyols possessing substantial functionality and low viscosity from 5 different vegetable oils such as canola oil (two grades), sunflower oil, flax oil and camalina oil was reported, followed by treatment with petrochemical derived diisocyanates to develop high-solid PU coatings exhibiting excellent mechanical and thermo-mechanical properties. In comparison, flax oil based PU had higher T_g and crosslinking density, low solvent swelling ratio, improved hydrophobic nature, excellent tensile properties, hardness and abrasion resistance over other vegetable oil based PUs [63]. Hygrothermal ageing of thermoplastic PU coatings was examined by Boubakri et al. [64] using PU coating samples immersed in water at 70 °C for 6 months. The authors evaluated water absorption as well as thermal, mechanical and tribological properties of the coatings. The coating samples exhibited high rate of Fickian water diffusion. A decrease in T_g value was observed owing to the plasticizing effect of coatings resulting from the absorption of water molecules. Ageing of PU coating samples was confirmed using FTIR results. Also, the ageing led to the reduction in the mechanical and tribological properties.

In another study, Rassoul et al. [65] investigated the cathodic protection of thermally insulated PU coated pipelines. The effect of specific gravity of PU and electrolytic (NaCl) concentration on the cathodic protection of coatings was studied. Samples with four distinct densities (35, 68, 86 and 113 kg/m^3) in four different electrolytic concentrations (0, 1.5, 3.5 and 5% NaCl) were examined at room and higher temperatures. PU coatings with density of 68 kg/m^3 in 3.5 and 5% NaCl concentrations were observed to have lower resistivity towards cathodic protection owing to inadequate PU resistivity and elevated electrolytic conductivity. Also, a current density of 30.9 mA/m^2 was observed to be suitable for the cathodic protection of PU with a density of 68 kg/m^3 in 3.5% NaCl solution after 1608 h of complete saturation. An adequate shielding potential of coatings to protect pipelines up to primary saturation of 240 h was observed. Furthermore, the coatings presented stabilized cathodic protection up to 60 °C. A study by Sousa et al. [66] evaluated corrosion under thermal insulation of rigid PU foams on the pipelines and the compatibility of the foams with anti-corrosive base coatings. For this purpose, the authors developed aqueous extracts of PU foams according to ASTM C871 and analyzed using chemical, electrochemical and mass loss characterizations. Substantial impact of temperature increase on pH, conductivity and contents of phosphate, chloride and fluoride was observed using chemical analysis. Also, the influence of other components in the PU foams besides flame retardant on under coating corrosion was revealed through the halides generation studies. Remarkable effect of chloride content in the PU foams on the corrosion was observed, which can be resolved only through proper control of chloride during PU foam preparation. Furthermore, the compatibility tests using two anti-corrosive coatings indicated that FBE was the best choice for new pipelines and a traditional maintenance coating (BAR RUST) for old pipeline repairs.

Motamedi *et al.* [67] reported the corrosion protection capability and adhesion strength of PU coatings on mild steel surfaces in the presence of 1 M sulfamic acid cleaning solution containing surfactants such as dodecyltrimethylammonium bromide (DTAB) and its counterpart 12-4-12. Enhanced adhesion strength and corrosion resistance behavior of PU coatings in both DTAB and 12-4-12 were detected. Acid cleaning solution containing DTAB presented slightly better performances when compared to 12-4-12. Low molecular weight of DTAB was advantageous to enhance the adsorption rate and, thereby, to enrich the performance. Akbarian *et al.* [68] evaluated the corrosion protection capability of nanoparticulate silver added PU coatings in 3.5% NaCl solution. The effect of silver nanoparticles on the corrosion performance of water based PU (WPU) and high solid PU (HPU) was analyzed using EIS, SEM and FTIR spectroscopy. The incorporation of silver nanoparticles had no remarkable effect on the corrosion protection of HPU, whereas caused coating deterioration in the case of WPU. Lesser amount of carbonyl groups and their decomposition under moisture environment attributed to the deterioration of silver nanoparticles impregnated WPU. Furthermore, silver nanoparticles offered antibacterial properties to the PU coating. Khun *et al.* [69] studied the cathodic delamination of PU composite coating containing multiwalled carbon nanotube (MWCNT) from the steel surfaces with the aid of scanning Kelvin probe (SKP). Increase in anti-corrosion properties as well as considerable decrease in cathodic delamination for PU/MWCNT composite coatings was observed with increase in MWCNT content from 0 to 0.5 wt%. Also, the composite coatings exhibited enhanced cathodic delamination resistance due to the increased barrier resistance against oxygen and water molecules resulting from uniform dispersion of highly dense MWCNT structures in PU matrix. In another study, the individual effect of graphene oxide (GO), mildy reduced graphene oxide (RGO) and functionalized graphene (FG) on the corrosion resistance properties of waterborne PU (WPU) was reported by Li et al. [70]. The corrosion behavior of composite coatings in 3.5% NaCl solutions was analyzed using EIS and salt spray tests. Enhanced corrosion resistance for all PU/graphene composites was observed. Good dispersion of GO and RGO in the PU matrix was observed, whereas a submicron sized aggregation was reported in the case of FG. Thus, the fine dispersion of GO and RGO offered outstanding corrosion barrier performances to PU over FG. PU composites containing 0.2 wt% RGO were reported to have optimal anti-corrosion properties and constant impedance modulus of 10^9 Ω at 0.1 Hz for up to 235 h [70].

1.3.4 Fluoropolymer Based Pipeline Coatings

Fluoropolymers are the fluorocarbon based polymers containing stronger carbon-fluorine bond in their chain structure. The history of this important class of polymeric materials dates back to 1938 with the accidental invention of polytetrafluoroethylene (PTFE or Teflon) by DuPont. Fluoropolymers have been utilized for a wide variety of coating applications because of their excellent performance characteristics. The unique properties of these polymers include high temperature stability, resistance to corrosion, abrasion and chemicals, high electrical resistivity, low surface energy and refractive index. However, fluoropolymers generally have poor solubility in conventional coating solvents and poor adhesion to metals and other substrates. To enhance the usage of fluoropolymers as coating materials, many developments have been reported to overcome the drawbacks. Generally, fluoropolymer coatings are used for the purposes of corrosion resistance as well as fouling-release in the oil/gas pipelines. Examples of fluoropolymers used for coating purposes comprise polytetrafluororoethylene (PTFE), tetrafluoroethylene/hexafluoroethylene copolymers (FEP), TFE/perfluoroalkyl vinyl ether copolymers (FEP), polyvinylidene fluoride (PVDF), fluoroethylene vinyl ether (FEVE), etc.

In a recent study, Lee et al. [71] reported transparent superhydrophobic and translucent superamphiphobic coatings via spraying silica–fluoropolymer hybrid nanoparticles (SFNs). Figure 1.8 demonstrates the fabrication process of the fluoropolymer modified silica nanoparticles. No specific modification of the substrate was required and the coatings could be applied to a variety of substrates. The modified silica nanoparticles exhibited strong potential of application in a wide variety of advanced coatings systems. In another

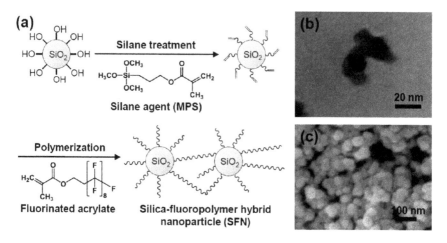

Figure 1.8 (a) Schematic of the fabrication process of fluoropolymer modified silica nanoparticles. TEM (b) and SEM (c) images of the hybrid nanoparticles (SFNs). Reproduced from Reference 71 with permission from American Chemical Society.

study reporting the enhancement of fluoropolymers for coatings applications, Gudipati et al. [72] developed crosslinked hyperbranched fluoropolymer (HBFP)-poly(ethylene glycol) (PEG) composite coatings (Figure 1.9). The coatings inhibited protein adsorption and marine organism settlement.

Darden et al. [73] discussed the modifications in the long-life fluoropolymer based coating (such as FEVE) to adjust with the environmental regulations. In this respect, new FEVE coating systems were developed, which included FEVE solid resins, FEVE water dispersions, FEVE water emulsions in blended resin system, silanol functional FEVE resins and fluorourethanes. Significant improvement in weathering resistance was observed for coatings based on FEVE blend. Fluorourethanes were also attractive for industrial use because of favourable life cycle. McKeen et al. [74] reported a sequence of new fluoropolymer based multilayered coating systems with outstanding adhesive strength for the inside surfaces of the oil and gas pipelines. Structurally, these coatings composed of blend of fluoropolymer and binder resins such as polyamide-imide, polyether sulfone or polyphenylene sulfide as primer, permeation barrier additives as midcoat and fluoropolymers as topcoat. The coatings were analyzed under various conditions of high temperature and

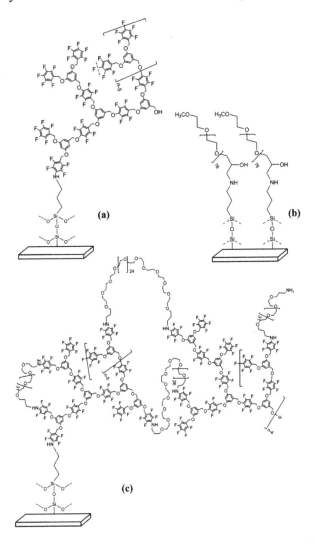

Figure 1.9 Representation of covalent attachment of (a) HBFP, (b) PEG, and (c) HBFP-PEG cross-linked networks to the substrate. Reproduced from Reference 72 with permission from American Chemical Society.

pressure in sweet, sour and hydrochloric acid environments. The properties of the coatings, especially adhesion strength, remained unchanged after exposure to variety of laboratory testing environments. Another study by McKeen et al. [75] generated a fluoropolymer based internal coating system for the protection of downhole production tubulars in the oil and gas wells. The coatings were examined using corrosion autoclave tests under salt and sweet environments at 8000 psi pressure and 325 °F temperature for 24 h and followed by a rapid decompression from 8000 psi to 1500 psi within 34 sec. In addition, the coatings were analyzed for abrasion resistance and deposition reduction. It was concluded from the study that the fluoropolymer based coatings were favored due to corrosion resistance at high temperatures and pressures as well as fouling-release against organic and inorganic depositions include asphaltene, paraffin and $BaSO_4$ scales [75].

Bayram *et al.* [76] studied the performance of perfluoroalkoxy (PFA) fluoropolymer and a hybrid epoxy-fluoropolymer resin for the corrosion protection of the interior surfaces of the oil and gas pipelines. The morphological, adhesion, hardness and corrosion resistance behavior of the coatings towards chlorine, hydrochloric acid and salt fog environments was investigated. PFA coatings were a three-layer system composed of pure PFA as top layer, blend of perfluoroethylene propylene (FEP) (59 wt%) and two non-fluoropolymer binders polyamide-imide (PAI) (5 wt%) and polyether sulfone (PES) (36 wt%) as primer and a blend of PFA and FEP as midcoat. In the case of PFA coatings, good adhesion of coatings with substrates was reported. Also, good interlayer adhesion and chemical resistance was observed under corrosive environments. PFA coatings along with primer and midcoat were confirmed to be suitable for corrosion resistance. Hybrid epoxy-fluoropolymer resin was a blend of diglycidyl ether of bisphenol A (DGEBA) and 10 wt% low molecular weight PTFE as single layer. The authors observed good adhesion strength of these hybrid coatings to the substrates, which was attributed to the incorporation of epoxy in the coating formulations. Further, the usage of PTFE provided high service temperature and corrosion resistance to the coatings.

Table 1.1 presents advantages, limitations and application of PE, PP, epoxies, PU and fluoropolymer based pipeline coating systems reported in various literature studies.

Table 1.1 Comparison of commonly used polymeric pipeline coatings, specifically their application

Polymer	Coating systems	Advantages	Limitations
Polyethylene	Polymeric tapes; Heat shrinkable tape coatings; 2-layer and 3-layer polyethylene coatings; High performance composite coatings (HPCC)	Good mechanical properties; Corrosion resistance; Chemical stability; Electrical resistance; Water and moisture resistance; Inexpensive	Inferior adhesion to metallic surfaces; Limited temperature range
Polypropylene	Polymeric tapes; Heat shrinkable sleeves; 2-layer and 3-layer polypropylene coatings	Good mechanical properties; Corrosion resistance; High temperature stability; Electrical resistance; Water and moisture resistance; Good chemical resistance; Cost-effective material	Inferior adhesion to metallic surfaces
Epoxies	Base coat primer; Fusion bonded epoxy (single and dual layer); Liquid coatings; Epoxy nanocomposites	Good adhesion; UV stability; Chemical resistance; Cathodic protection; Stress cracking resistance; Resistance to biological environments; Improved fluid flow characteristics	Low curing speed
Polyurethanes	Liquid coatings; Thermoplastic PU; 100% solid PU; Rigid PU foams; PU nanocomposites	Good corrosion resistance; Optimal chemical and solvent resistance; Tear propagation resistance; High environmental degradation re-	Low moisture resistance; Complexity in application

		sistance; Superior abrasion and scratch resistance; Excellent thermal resistance; Low curing time and high curing density	
Fluoropolymers	Solid resins; Water dispersions; Emulsions in blended resin system; Silanol functionalized system; Fluorourethanes; Multilayered coating system; Hybrid fluoro-resins	Corrosion resistance; High temperature stability; Good abrasion and chemical resistance; Low surface energy and refractive index	Inferior adhesion to metallic surfaces; Poor solubility in conventional coating solvents

1.4 Conclusions

In this chapter, polymer coatings used for the pipeline protection in oil and gas industries have been reviewed, progressing from epoxy primer of four components coal tar enamel (CTE) system in the early stages to the next generation polymer nanocomposite based coatings under development today. Polymer based coatings provide effective solutions for external as well as internal protection of pipelines from corrosion, fouling and mechanical damage. Of the various polymeric materials, polyethylene, polypropylene, epoxies, polyurethanes and fluoropolymers have attractive property profiles. Polyolefin based coatings have excellent mechanical properties, corrosion, moisture and chemical resistance, health and safety adherence and low cost. Performance characteristics of both polyethylene and polypropylene based coatings have gained remarkable developments over the last decades so as to achieve suitability for application in challenging environments. Protective coatings based on epoxies either as stand-alone liquid/powder or as part of other coating systems also exhibit good performance profiles, including good adhesion strength to metallic surfaces, cathodic protection and resistance to chemicals, stress cracking and biological environments.

Similarly, PU pipeline coatings have good adhesion, low temperature impact strength, adherence to health and safety regulations, fast curing speed, thermal insulation and resistance to corrosion, abrasion, scratch and tear propagation. Fluoropolymer based coating systems have also acquired wide acceptance and usefulness for pipeline protection. These polymers offer outstanding corrosion resistance at elevated temperatures and pressures, along with good fouling release owing to their unique features such as high temperature stability, low surface energy and refractive index, resistant to corrosion, abrasion and chemicals, high electrical resistivity, etc. Thus, polymer materials enhance the life span of oil and gas pipelines through either corrosion control or rehabilitation of corroded parts as protective coating materials.

References

1. Banach, J. L. (1987) Pipeline Coatings - Evaluation, Repair, and Impact on Corrosion Protection Design and Cost. *Corrosion '87*, USA, CONF-870314.
2. Rueda, F., Otegui, J. L., and Frontini, P. (2012) Numerical tool to model collapse of polymeric liners in pipelines. *Engineering Failure Analysis*, **20**, 25-34.
3. Yang, X.-H., Zhu, W.-L., Lin, Z., and Huo, J.-J. (2005) Aerodynamic evaluation of an internal epoxy coating in nature gas pipeline. *Progress in Organic Coatings*, **54**, 73-77.
4. McGill, J. C., McGill, J. M., and Key, B. L. (1999) Pipeline Coating, US Patent 5984581 A.
5. Samour, C. M., Jackson, E. G., Thomas, S. J., and Davidson, L. E. (1980) Coated Pipe and Process for Making Same, US Patent 4213486 A.
6. Guermazi, N., Elleuch, K., and Ayedi, H. (2009) The effect of time and aging temperature on structural and mechanical properties of pipeline coating. *Materials & Design*, **30**, 2006-2010.
7. Papavinasam, S., Attard, M., Balducci, B., and Revie, R. W. (2009) Testing coatings for pipeline: new laboratory methodologies to simulate field operating conditions of external pipeline coatings. *Journal of Protective Coatings & Linings*, 32-51.
8. Papavinasam, S., Attard, M., and Revie, R. W. (2006) External polymeric pipeline coating failure modes. *Materials Performance*, **45**, 28-30.
9. Samimi, A., and Zarinabadi, S. (2011) An analysis of polyethylene coating corrosion in oil and gas pipelines. *Journal of American Science*, **7**, 1032-1036.

10. Samimi, A., Dokhani, S., Neshat, N., Almasinia, B., and Setoudeh, M. (2012) The application and new mechanism of universal produce the 3-layer polyethylene coating. *International Journal of Advanced Scientific and Technical Research,* **2**, 465-473.
11. Romano, M., Dabiri, M., and Kehr, A. (2005) The ins and outs of pipeline coatings: Coatings used to protect oil and gas pipelines. *Journal of Protective Coatings & Linings,* **22**, 40-47.
12. Sloan, R. N. (2001) Pipeline coatings. In: *Control of Pipeline Corrosion,* Peabody, A. W., and Bianchetti, R. L. (eds.), 2nd edition, NACE International, USA, pp. 7-20.
13. Niu, L., and Cheng, Y. (2008) Development of innovative coating technology for pipeline operation crossing the permafrost terrain. *Construction and Building Materials,* **22**, 417-422.
14. Jack, T. R., Wilmott, M. J., Sutherby, R. L., and Worthingham, R. G. (1996) External corrosion of line pipe - A summary of research activities. *Materials Performance,* **35**, 18-24.
15. Roche, M., and Melot, D. (2012) Recent experience with pipeline coating failures. In: *Protecting and Maintaining Transmission Pipe,* Technology Publishing Company, pp. 57-66. Online: http://www.paintsquare.com/store/assets/JPCL_transpipe_ebook.pdf#page=62 (assessed 18th February 2017).
16. Kehr, J. A. (2012) How fusion-bonded epoxies protect pipeline: single-and double-layer systems. In: *Protecting and Maintaining Transmission Pipe,* Technology Publishing Company, pp. 13-22. Online: http://www.paintsquare.com/store/assets/JPCL_transpipe_ebook.pdf#page=62 (assessed 18th February 2017).
17. Enos, D., Kehr, J., and Guilbert, C. A high-performance, damage-tolerant, fusion-bonded epoxy coating. Online: alankehr-anti-corrosion.com (assessed 21st February 2017).
18. Kehr, J., Hislop, R., Anzalone, P., and Kataev, A. (2012) Liquid coatings for girthwelds and joints: proven corrosion protection for pipelines. In: *Protecting and Maintaining Transmission Pipe,* Technology Publishing Company, pp. 23-32. Online: http://www.paintsquare.com/store/assets/JPCL_transpipe_ebook.pdf#page=62 (assessed 18th February 2017).
19. Goldie, B. (2010) Developments in pipeline protection reviewed. *Journal of Protective Coatings & Linings,* 30-32.
20. Luckachan, G. E., and Mittal, V. (2015) Anti-corrosion behavior of layer by layer coatings of cross-linked chitosan and poly(vinyl butyral) on carbon steel. *Cellulose,* **22**, 3275-3290.
21. Soares, J. B., and McKenna, T. F. (2013) *Polyolefin Reaction Engineering,* John Wiley & Sons, USA.
22. Samimi, A., and Zarinabadi, S. (2012) Application of polyurethane as coating in oil and gas pipelines. *International Journal of Science*

and *Investigations,* **1**, 43-45.
23. Roche, M. G. (2004) An experience in offshore pipeline coatings. *Corrosion 2004,* USA. Online: https://www.onepetro.org/conference-paper/NACE-04018 (assessed 28th February 2017).
24. Stafford, T. (1998) *Plastics in Pressure Pipes, Report 102,* Rapra Technology Limited, UK.
25. Soucek, M. D. (2012) The application and new mechanism of universal produce the 3-layer polyethylene coating," *International Journal of Chemistry,* **1**, 94-104.
26. Chang, B., Sue, H.-J., Wong, D., Kehr, A., Pham, H., Siegmund, A., Snider, W., Jiang, H., Browning, B., and Mallozzi, M. (2008) Integrity of 3LPE Pipeline Coatings – Residual Stresses and Adhesion Degradation. *7th International Pipeline Conference,* Canada, pp. 75-86.
27. Kamimura, T., and Kishikawa, H. (1998) Mechanism of cathodic disbonding of three-layer polyethylene-coated steel pipe. *Corrosion,* **54**, 979-987.
28. Guermazi, N., Elleuch, K., Ayedi, H., and Kapsa, P. (2008) Aging effect on thermal, mechanical and tribological behaviour of polymeric coatings used for pipeline application. *Journal of Materials Processing Technology,* **203**, 404-410.
29. Guermazi, N., Elleuch, K., Ayedi, H., Fridrici, V., and Kapsa, P. (2009) Tribological behaviour of pipe coating in dry sliding contact with steel. *Materials & Design,* **30**, 3094-3104.
30. Guermazi, N., Elleuch, K., Ayedi, H., Zahouani, H., and Kapsa, P. (2008) Susceptibility to scratch damage of high density polyethylene coating. *Materials Science and Engineering A,* **492**, 400-406.
31. Samimi, A. (2012) Study an analysis and suggest new mechanism of 3 layer polyethylene coating corrosion cooling water pipeline in oil refinery in Iran. *International Journal of Innovation and Applied Studies,* **1**, 216-225.
32. Guan, S. W., Gritis, N., Jackson, A., and Singh, P. (2005) Advanced Onshore and Offshore Pipeline Coating Technologies. *2005 China International Oil & GasPipeline Technology (Integrity) Conference & Expo,* China. Online: http://citeseerx.ist.psu.edu/viewdoc/download?doi=10.1.1.131.4370&rep=rep1&type=pdf (assessed 21st February 2017).
33. Howell, G., and Cheng, Y. (2007) Characterization of high performance composite coating for the northern pipeline application. *Progress in Organic Coatings,* **60**, 148-152.
34. Singh, P. J., and Cox, J. J. (2000) Development of a Cost-effective Powder Coated Multi-component Coating for Pipelines. *Corrosion 2000,* USA. Online: https://www.onepetro.org/conference-paper/NACE-00762 (assessed 1st March 2017).
35. Lam, C., Wong, D. T., Steele, R., and Edmondson, S. (2007) A New

Approach to High Performance Polyolefin Coatings. *Corrosion 2007*, USA. Online: http://w.brederoshaw.com/non_html/techpapers/BrederoShaw_TP_G_04.pdf (asessed 21st February 2017).

36. Karger-Kocsis, J. (2012) *Polypropylene Structure, Blends and Composites: Volume 3, Composites*, Springer Science & Business Media, USA.
37. Fairhurst, D., and Willis, D. (997) Polypropylene coating systems for pipelines operating at elevated temperatures. *Journal of Protective Coatings & Linings*, **14**, 64.
38. Suzuki, K., Ishida, M., Ohtsuki, F., Inuizawa, Y., Hinenoya, S., Tanaka, M., and Shindou, Y. (1986) Polypropylene Coated Steel Pipe, US Patent 4606953 A.
39. Guidetti, G., Rigosi, G., and Marzola, R. (1996) The use of polypropylene in pipeline coatings. *Progress in Organic Coatings*, **27**, 79-85.
40. Moosavi, A. N., Al-Mutawwa, S. O., Balboul, S., and Saady, M. R. (2006) Hidden Problems with Three Layer Polypropylene Pipeline Coatings. *Corrosion 2006*, USA. Online: https://www.onepetro.org/conference-paper/NACE-06057 (assessed 25th February 2017).
41. Moosavi, A. N., Chang, B. T., and Morsi, K. (2010) Failure Analysis Of Three Layer Polypropylene Pipeline Coatings. *Corrosion 2010*, USA. Online: https://www.onepetro.org/conference-paper/NACE-10002 (assessed 19th February 2017).
42. Pascault, J.-P., and Williams, R. J. (2009) *Epoxy Polymers*, John Wiley & Sons, USA.
43. Kehr, J. A., and Enos, D. G. (2000) FBE, A Foundation for Pipeline Corrosion Coatings. *Corrosion 2000*, USA. Online: https://www.onepetro.org/conference-paper/NACE-00757 (assessed 25th February 2017).
44. Wei, Y., Zhang, L., and Ke, W. (2006) Comparison of the degradation behaviour of fusion-bonded epoxy powder coating systems under flowing and static immersion. *Corrosion Science*, **48**, 1449-1461.
45. Luo, X., and Mather, P. T. (2013) Shape memory assisted self-healing coating. *ACS Macro Letters*, **2**, 152-156.
46. Augustyniak, A., Tsavalas, J., and Ming, W. (2009) Early detection of steel corrosion via "turn on" fluorescence in smart epoxy coatings. *ACS Applied Materials and Interfaces*, **1**, 2618-2623.
47. Wei, Y., Zhang, L., and Ke, W. (2007) Evaluation of corrosion protection of carbon black filled fusion-bonded epoxy coatings on mild steel during exposure to a quiescent 3% NaCl solution. *Corrosion Science*, **49**, 287-302.
48. Goertzen, W. K., and Kessler, M. (2006) Creep behavior of carbon

fiber/epoxy matrix composites. *Materials Science and Engineering A,* **421**, 217-225.

49. Alamilla, J., Sosa, E., Sanchez-Magana, C., Andrade-Valencia, R., and Contreras, A. (2013) Failure analysis and mechanical performance of an oil pipeline. *Materials & Design,* **50**, 766-773.
50. Zhou, W., Jeffers, T. E., and Decker, O. H. (2007) Properties of a novel high T_g FBE coating for high temperature service. *Materials Performance,* **46**, 36-40.
51. Moon, B., Paek, B., Lee, J., Wang, H., and Lee, N. (2011) New Development of a High T_g FBE Coating. *Corrosion 2011*, USA. Online: https://www.onepetro.org/conference-paper/NACE-11032 (assessed 23rd February 2017).
52. Duell, J., Wilson, J., and Kessler, M. (2008) Analysis of a carbon composite overwrap pipeline repair system. *International Journal of Pressure Vessels and Piping,* **85**, 782-788.
53. Yuan, Y. C., Rong, M. Z., Zhang, M. Q., Chen, J., Yang, G. C., and Li, X. M. (2008) Self-healing polymeric materials using epoxy/mercaptan as the healant. *Mcromolecules,* **41**, 5197-5202.
54. Bakhshandeh, E., Jannesari, A., Ranjbar, Z., Sobhani, S., and Saeb, M. R. (2014) Anti-corrosion hybrid coatings based on epoxy–silica nano-composites: Toward relationship between the morphology and EIS data. *Progress in Organic Coatings,* **77**, 1169-1183.
55. Weng, C.-J., Chang, C.-H., Peng, C.-W., Chen, S.-W., Yeh, J.-M., Hsu, C.-L., and Wei, Y. (2011) Advanced anticorrosive coatings prepared from the mimicked xanthosoma sagittifolium-leaf-like electroactive epoxy with synergistic effects of superhydrophobicity and redox catalytic capability. *Chemistry of Materials,* **3**, 2075-2083.
56. Rosu, D., Rosu, L., and Cascaval, C. N. (2009) IR-change and yellowing of polyurethane as a result of UV irradiation. *Polymer Degradation and Stability,* **94**, 591-596.
57. Engels, H. W., Pirkl, H. G., Albers, R., Albach, R. W., Krause, J., Hoffmann, A., Casselmann, H., and Dormish, J. (2013) Polyurethanes: Versatile materials and sustainable problem solvers for today's challenges. *Angewandte Chemie International Edition,* **52**, 9422-9441.
58. Guan, S. W. (2003) 100% Solids Rigid Polyurethane Coatings Technology and Its Application on Pipeline Corrosion Protection. *Pipeline Engineering and Construction International Conference 2003.* Online: http://ascelibrary.org/doi/pdf/10.1061/40690%282003%2928#sthash.5BKyd9zK.dpuf (assessed 1st March 2017).
59. Ghosal, A., Rahman, O. U., and Ahmad, S. (2015) High-performance soya polyurethane networked silica hybrid nanocomposite coatings. *Industrial and Engineering Chemistry Research,* **54**, 12770-12787.

60. Chattopadhyay, D. K., Sreedhar, B., and Raju, K. V. S. N. (2005) Effect of chain extender on phase mixing and coating properties of polyurethane ureas. *Industrial and Engineering Chemistry Research*, **44**, 1772-1779.
61. Samimi, A. (2012) Use of polyurethane coating to prevent corrosion in oil and gas pipelines transfer. *International Journal of Innovation and Applied Studies*, **1**, 186-193.
62. Guan, S. W. (2003) Advanced 100% Solids Rigid Polyurethane Coatings Technology for Pipeline Field Joints and Rehabilitation. *Corrosion 2003*, USA. Online: http://www.penderlo.com/doc/PUTech.pdf (assessed 2nd March 2017).
63. Kong, X., Liu, G., Qi, H., and Curtis, J. M. (2013) Preparation and characterization of high-solid polyurethane coating systems based on vegetable oil derived polyols. *Progress in Organic Coatings*, **76**, 1151-1160.
64. Boubakri, A., Elleuch, K., Guermazi, N., and Ayedi, H. (2009) Investigations on hygrothermal aging of thermoplastic polyurethane material. *Materials & Design*, **30**, 3958-3965.
65. Rassoul, E.-S. A., Abdel-Samad, A., and El-Naqier, R. (2009) On the cathodic protection of thermally insulated pipelines. *Engineering Failure Analysis*, **16**, 2047-2053.
66. De Sousa, F., Da Mota, R., Quintela, J., Vieira, M., Margarit, I., and Mattos, O. (2007) Characterization of corrosive agents in polyurethane foams for thermal insulation of pipelines. *Electrochimica Acta*, **52**, 7780-7785.
67. Motamedi, M., Tehrani-Bagha, A., and Mahdavian, M. (2014) The effect of cationic surfactants in acid cleaning solutions on protective performance and adhesion strength of the subsequent polyurethane coating. *Progress in Organic Coatings*, **77**, 712-718.
68. Akbarian, M., Olya, M., Mahdavian, M., and Ataeefard, M. (2014) Effects of nanoparticulate silver on the corrosion protection performance of polyurethane coatings on mild steel in sodium chloride solution. *Progress in Organic Coatings*, **77**, 1233-1240.
69. Khun, N., and Frankel, G. (2016) Cathodic delamination of polyurethane/multiwalled carbon nanotube composite coatings from steel substrates. *Progress in Organic Coatings*, **99**, 55-60.
70. Li, J., Cui, J., Yang, J., Li, Y., Qiu, H., and Yang, J. (2016) Reinforcement of graphene and its derivatives on the anticorrosive properties of waterborne polyurethane coatings. *Composites Science and Technology*, **129**, 30-37.
71. Lee, S. G., Ham, D. S., Lee, D. Y., Bong, H., and Cho, K. (2013) Transparent superhydrophobic/translucent superamphiphobic coatings based on silica–fluoropolymer hybrid nanoparticles. *Langmuir*, **9**, 15051-15057.

72. Gudipati, C. S., Finlay, J. A., Callow, J. A., Callow, M. E., and Wooley, K. E. (2005) The antifouling and fouling-release perfomance of hyperbranched fluoropolymer (HBFP)–poly(ethylene glycol) (PEG) composite coatings evaluated by adsorption of biomacromolecules and the green fouling alga Ulva. *Langmuir*, **1**, 3044-3053.
73. Darden, W., and Parker, B. (2006) Advances in Fluoropolymer Resins for Long-life Coatings. *PACE*, USA. Online: http://lumiflonusa.com/wp-content/uploads/2010_PACE_Paper.pdf (assessed 5th March 2017).
74. Albert, R., McKeen, L. W., and Hofmans, J. (2008) Corrosion Testing of Fluorocoatings for Oil/Gas Production Tubing. *Corrosion 2008*, USA. Online: https://www.onepetro.org/conference-paper/NACE-08029 (assessed 26th February 2017).
75. Hofmans, J., Nelissen, J., Tixhon, J.-M., and McKeen, L. W. (2012) Engineered Internal Downhole Coating Solutions for Corrosion & Deposition Control in Downhole Production Tubulars. *Abu Dhabi International Petroleum Conference and Exhibition*, UAE. Online: https://www.onepetro.org/conference-paper/SPE-161204-MS (assessed 2nd March 2017).
76. Bayram, T. C., Orbey, N., Adhikari, R. Y., and Tuominen, M. (2015) FP-based formulations as protective coatings in oil/gas pipelines. *Progress in Organic Coatings*, 88, 54-63.

2

Ionic Liquids and Poly(ionic liquid)s Based Coatings

2.1 Introduction

The successful story of ionic liquids (ILs) commenced with the introduction of air and water stable 1-ethyl-3-methylimidazolium based IL in an interesting study by Wilkes and Zaworotko [1]. Following this, significant research efforts have been undertaken to enhance the performance of ILs and, thus, applicability range. ILs are high performance environmentally sustainable ambient temperature organic liquids of salts basically composed of an organic cation and a molecular anion with either inorganic or organic nature [2-5]. Due to the ever-increasing environmental concerns, the IL based materials have received further research attention due to their potential to replace conventional volatile chemicals used to generate solutions of molecules and to design functional structures, supports, systems and fluids. The benefits of IL based materials enable their suitability in diverse application areas including catalytic and organic reactions, coatings, energy storage and conversion, extraction, sorption, electrolysis and pharmaceutical [6-11]. Specifically, the distinctive features of ILs driving their potential applications include modulating structural as well as performance profile, outstanding flow resistance, appreciable thermal, flame and chemical reliability, stabilized electrochemical features, nominal volatility, superior solubilizing power, innocuous nature, etc. (Figure 2.1) [11-19].

Poly(ionic liquid)s (PILs), also known as polymerizable ionic liquids, are a class of specialty polymers under the category of polyelectrolytes displaying synergistic features of both ILs and macromolecules, which include stabilized electrochemical and thermal characteristics of ILs and considerable mechanical and dimensional stability of macromolecules [20,21]. As a result, their application potential in diverse industrial fields is significantly enhanced. Generally, poly(ionic liquid)s are considered as advanced analogs of ionic liquids, developed for attenuating the common limitations of ionic

Anish M. Varghese and Vikas Mittal, Khalifa University of Science and Technology, Abu Dhabi, UAE
© 2021 Central West Publishing, Australia

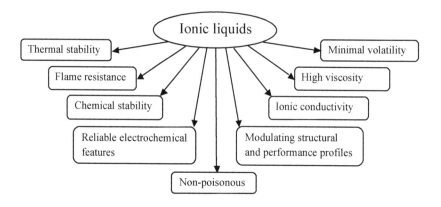

Figure 2.1 Properties of ionic liquids.

liquids such as discharge issues, miniaturization impracticability due to liquid form, etc. [22,23]. These high quality polymeric materials can be commonly designed with the aid of addition or condensation polymerization reactions of corresponding IL monomeric units or by direct IL induced chemical functionalization of neat macromolecules [24,25]. PILs are solid structures comprising of both positively and negatively charged ions and macromolecular unit with low glass transition temperature (T_g) values [26,27]. So far, macromolecules of various monomers like styrene, methyl methacrylate, ethylene oxide, ethylene glycol, vinyl esters, esters, imides, siloxane and ionenes have demonstrated practicability for the beneficial generation of PILs [28-36]. Also, the positively charged ions such as pyrrolidinium, pyridinium, imidazolium, alkyl-phosphonium and ammonium, guanidinium, etc. and the negatively charged ions such as hexafluorophosphate, tetrafluoroborate, tricyanomethanide, triflates, dicyanamide, bis(fluorosulfonyl)imide, etc., have been widely utilized as ionic moieties for the synthesis of PILs [24,25,37,38]. The molecular arrangement of PILs is either linear or branched. The branched architecture exhibits enhanced material and chemical features. Some branched PILs and oligomeric ILs are also demonstrated in Figure 2.2 [39].

Considering the advantages of ILs and PILs, these materials are widely studied to design functional coatings for diverse engineering applications. To highlight it further, the present chapter summarizes the recent literature studies presenting various advancements in the development of different IL and PIL based coatings.

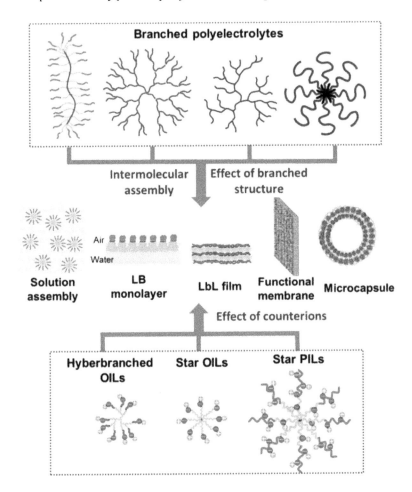

Figure 2.2 Top: structure of branched polyelectrolytes; bottom: structure of major types of branched PILs and oligomeric ILs; middle: representative examples of assembled structures. Reproduced from Reference 39 with permission from American Chemical Society.

2.2 Poly(ionic liquid)s Based Coatings

2.2.1 Poly(ionic liquid)s Based Protective Coatings

Owing to the combined properties of ILs and macromolecules, PILs have been used for the fabrication of high performance protective coatings with minimal environmental impact during the processing and applications stages. Taghavikish *et al.* [40] detailed the development of corrosion resistant coating system based on interconnected PIL nano-emulsion. For this purpose, light induced thiol-ene polymerization was carried out with the assistance of 1,4-di(vinylimidazolium) butane bisbromide (DVIMBr) and pentaerythritol tetrakis(3-mercaptopropionate) (PETKMP), along with other additives such as initiator, surfactant and diluent. The resultant PIL nano-emulsion was utilized as an inhibitor in a self-assembled nanoparticle (SNAP) coating and exhibited outstanding resistance against corrosion, which was confirmed with the aid of both electrochemical impedance spectroscopy (EIS) analysis and potentiodynamic polarization technique. In another study, the authors reported the fabrication of PIL gel beads exhibiting self-healing quality by employing the same technique and precursors (Figure 2.3) [41].

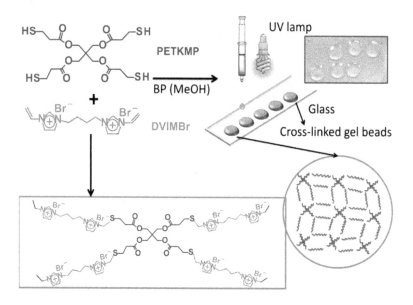

Figure 2.3 Schematic representation of the crosslinked PIL gel beads generated via light induced thiol-ene polymerization. Reproduced from Reference 41 with permission from American Chemical Society.

In order to ensure effective barrier features and, thereby, to offer anti-corrosion properties for metals, Taghavikish *et al.* [42] also reported the successful application of a layered coating system consisting of a PIL composite and a self-assembled nanophase particle (SNAP) coating. In this system, the generation of PIL composite based coating was achieved through the effective exploitation of 3-mercaptopropyl-trimethoxysilane and DVIMBr by means of light induced thiol-ene polymerization and successive sol-gel process. Significant thermal stability and high T_g value, together with effective corrosion resistance were noted for the composite coating due to the evolution of strong binary crosslinked structure.

2.2.2 Poly(ionic liquid)s Based Sorbent Coatings

Coatings based on different PILs have been developed as solid-phase microextraction materials owing to their sorption characteristics and environmental friendly character. A study by Cagliero et al. [43] reported the use of two PIL solid-phase microextraction (SPME) fibers possessing thin sorbent coatings, namely, 50% (w/w) 1,12-di(3-vinylimidazolium)dodecane dibromide [(VIM)$_2$C$_{12}$] 2[Br] in 1-vinyl-3-hexylimidazolium chloride [VHIM][Cl] (fiber 1) and 50% [(VIM)$_2$C$_{12}$] 2[NTf$_2$] in 1-vinylbenzyl-3-hexadecylimidazolium bis[(trifluoromethyl)sulfonyl]imide [VBHDIM][NTf$_2$] (fiber 2) with the assistance of spin coating technique. The authors studied the sensitivity of the crosslinked PILs based coatings for the microextraction of acrylamide present in the brewed coffee and coffee powder. The PILs based sorbent coatings demonstrated superior acrylamide sensitivity when compared to commonly used coating materials. In another study reporting the structured coatings, the authors observed that the PIL fiber based sorbent coating consisting of 50% 1,12-di(3-vinylbenzylbenzimidazolium)dodecane dibis[(trifluoromethyl)sulfonyl]imide as IL crosslinker in 1-vinyl-3-(10-hydroxydecyl)imidazolium bis[(trifluoromethyl)sulfonyl]imide IL monomer resulted in a limit of quantitation of 0.5µgL^{-1} [44]. Due to the occupancy of OH functionality in the IL cation, increased sensitivity of the PIL toward acrylamide was observed. Ho et al. [45] also disclosed the development of PILs based SPME crosslinked sorption coatings with the aid of monocationic ILs via light induced polymerization. The resultant sorbent coatings exhibited superior phthalate ester selectivity from brewed coffee.

With an objective to extract polar organic contaminants from water molecules, Pacheco-Fernandez et al. [46] reported the application of PILs based SMPE sorbent coatings. Superior contaminant selectivity for all studied crosslinked PILs based sorbent coatings was observed, when compared to commercially available grades. Coatings based on PILs of vinylbenzyl or vinyl alkyl imidazolium displayed significant potential for use in efficient water contaminant sorption together with superior solvent permanence and cyclability. In another related study, Cordero-Vaca et al. [47] also studied microextraction of the crosslinked PIL coatings based on vinyl-alkylimidazolium- (ViCnIm-) or vinylbenzyl-alkylimidazolium- (ViBzCnIm-) IL monomers, and di-(vinylimidazolium)dodecane ((ViIm)2C12-) or di-(vinylbenzylimidazolium)dodecane ((ViBzIm)2C12-) dicationic IL crosslinkers with the help of gas chromatography. Limit of detection value as low as 135 µgL^{-1} for *p*-cresol was observed.

For the removal of noxious aromatic compounds from cattle milk and water, Joshi et al. [48] implemented SPME using network structured PIL coating. The resultant PIL based SPME coating demonstrated improved sensitivity and limit of detection in ng/L range towards polychlorinated biphenyls compared to commonly available siloxane based coatings owing to the occurrence of aromatic groups in the crosslinked molecular structure. In another study, Zhang et al. [49] detailed the design of SPME coating with superior sorption capability using interconnected gelatinous composite structure prepared by copolymerization of 1-vinyl-3-butylimidazolium bis[(trifluoromethyl)sulfonyl]imide ([VC$_4$IM][NTf$_2$]) with the ionic liquid cross-linker, 1,12-di(3-vinylimidazolium)dodecane bis[(trifluoromethyl)sulfonyl]imide ([(VIM)$_2$C$_{12}$] 2[NTf$_2$]), along with multiwall carbon nanotubes (MWCNTs). The composite coating system displayed high sensitivity and withdrawing tendency towards noxious aromatic compounds in water due to the generation of enriched interaction between them. Also, the performance of the coating system could be modulated as per the need and the one containing higher amount of MWCNTs displayed substantial partition coefficient.

Trujillo-Rodriguez et al. [50] compared the performance of various PILs based sorbent coatings for the determination of volatile substances contained in cheese. It was observed that the PIL materials based on 1-vinyl-3-hexylimidazolium chloride IL monomer and 1,12-di(3-vinylimiazolium)dodecane dibromide IL crosslinker exhibited superior performance in extraction of the analytes as compared to other examined PILs as well as commercially available sorbent coating grades. Furthermore, the PIL coatings generated with the assistance of light induced polymerization technique presented higher extraction properties than the coatings generated by free-radical polymerization technique. Yu et al. [51] discussed the practicability of various PILs based crosslinked coating structures possessing substantial sorption proficiency for disparate medicines, pesticides, and phenolics. The PILs based coatings demonstrated considerable permanence in acidic and other solvent containing mediums. Among various PILs, PIL based on IL monomer 1-vinyl-3-(10-hydroxydecyl) imidazolium chloride and IL crosslinker 1,12-di(3-vinylbenzylimidazolium) dodecane dichloride exhibited greater selectivity, along with an improvement in selectivity degree with corresponding rise in sorbent coating thickness.

Jia et al. [52] reported sorbent composite coating system consisting of crosslinked PIL based on 1-vinyl-3-nonanol imidazolium bromide on titanium wire possessing substantial porosity at nanoscale level. Improved aliphatic alcoholic selectivity as well as extraction potential was observed for the PIL composite coating compared to neat PIL and titanium wire. In another study, Chen et al. [53] also disclosed the possibility of PIL coated (based on 1-vinyl-3-hexylimidazolium hexafluorophosphate as monomer and methylacryloyl-substituted polyhedral oligomeric silsesquioxane (POSS) as cross-linker) titanium wire for the robust SPME of perfluorinated substances. Table 2.1 illustrates the summary of PILs based sorbent coatings studied for various applications.

2.2.3 Other Functional Coatings Based on PILs

PIL based coatings also find applicability for other applications such as antistatic and optically clear coatings. As highlighted by Roessler et al. [54], the introduction of series of ILs based on 1-allyl-3-alkylimidazolium was beneficial to design wood flooring coating exhibiting superior antistatic proficiency. Among various photopolymerized ILs coatings, 1-allyl-3-methylimidazolium chloride IL demonstrated optimal antistatic performance. Also, leaching resistance and transparency was observed for the coating. Iwata et al. [55] also described the implementation of bis(trifluoromethanesulfonyl)imide based ILs in the polyurethane coating formulation to provide sustainable antistatic capability.

Isik et al. [56] reported the fabrication of clear coatings composed of composite of cellulose with cholinium lactate methacrylate IL monomeric units by means of photoinitiated polymerization (Figure 2.4). Superior optical clarity of the composite coating system resulted due to the high extent of cellulose dissolution in the ionic liquid [57].

Table 2.1 Summary of PIL based sorbent coatings studied for various applications

Authors	SPME purpose	Remarks
Cagliero et al. [43]	Acrylamide extraction from coffee powder	Limit of detection to low value
Ho et al. [45]	Phthalate ester extraction from brewed coffee	Superior selectivity
Pacheco-Fernandez et al. [46]	Extraction of polar organic contaminants from water	High extraction capability, solvent stability and cyclability
Cordero-Vaca et al. [47]	Extraction of water contaminants	Limit of detection to low value
Joshi et al. [48]	Removal of noxious aromatic compounds from cattle milk and water	Improved sensitivity and low value detection limit
Zhang et al. [49]	Removal of noxious aromatic compounds from water	Superior extraction capability and selectivity
Trujillo-Rodriguez et al. [50]	Volatile substances from cheese	Superior selectivity and extraction potential
Yu et al. [51]	Extraction of disparate medicines, pesticides, and phenolics	Great extraction potential and selectivity
Jia et al. [52]	Extraction of aliphatic alcohols	Improved extraction capability and selectivity as compared to individual components
Chen et al. [53]	Extraction of perfluorinated substances	Good extraction capability

Tokuda et al. [58] described the emulsion synthesis of PIL coating using 2-(methacryloyloxy)ethyl-trimethylammonium bis(trifluoromethanesulfonyl)amide monomeric unit. The coating exhibited swappable water repelling and water absorption behavior in the presence of air and water. Figure 2.5 demonstrates this behavior in comparison with polyHEMA film in terms of the water contact angle values. The contact angle of water on the PIL emulsion film was ~70°, which indicated hydrophobic surface. The retreating contact angle of around ~28° also exhibited water wettability of the surface. However, the surface was observed to be hydrophobic again

Ionic liquids and Poly(ionic liquid)s Based Coatings 45

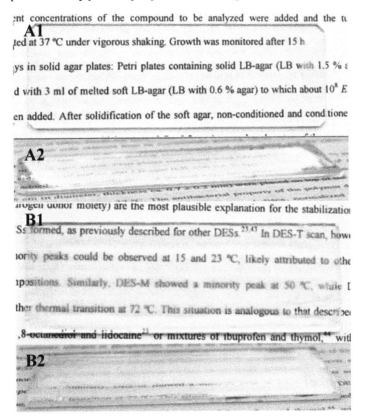

Figure 2.4 Cellulose-poly(2-cholinium lactate methacrylate) biocomposite coatings with (A) 5 wt % and (B) 10 wt % of cellulose. Reproduced from Reference 56 with permission from American Chemical Society.

when complete retreat of water had taken place.

2.3 Ionic Liquids Based Coatings

2.3.1 Ionic Liquids Based Coatings for Titanium

In order to enrich the surface behavior of titanium based materials, especially for biomedical purposes, IL coatings find feasibility owing to their excellent properties, biocompatibility, non-toxicity, etc. Gindri *et al.* [59] reported the application of IL coatings comprising of dicationic imidazolium for modifying the surface behavior of titanium. The resulting IL based coatings demonstrated superior stability and biocompatibility with the human gingival fibroblasts (HGF-1) as

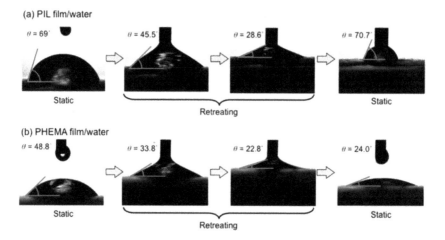

Figure 2.5 Water contact angle on the (a) PIL emulsion film and (b) poly(2-hydroxyethyl methacrylate) (polyHEMA) film. Reproduced from Reference 58 with permission from American Chemical Society.

well as pre-osteoblast (MC3T3-E1) cells. Also, the coatings exhibited outstanding anti-bacterial and anti-biofilm activity. In another study, the authors examined the effect of the structure of dicationic imidazolium-based ionic liquids (ILs) having bis(trifluoromethylsulfonyl)imide (NTf_2) and amino acid–based (methionine and phenylalanine) anionic moieties on the extent of adherence with titanium [60]. The authors confirmed that the factors like concentration of charge on IL moieties, presence of functional groups (like carboxylate and amino groups) leading to surface interactions via hydrogen bonding, etc., influenced the deposition profile of the coatings (Figure 2.6). In another study, Frizzo et al. [61] reported the development of docusate and iboprofenate based ILs possessing innocuous nature and microbial resistance suitable for generating high performance titanium coatings in order to widen the biomedical applications.

Siddiqui et al. [62] reported IL coating of dicationic imidazolium for the surface modification of titanium alloy. The IL coatings containing amino acid and phenylalanine based anionic species on the alloy surface demonstrated outstanding stability, improved anti-corrosion performance and appreciable wear resistance. In another study, Gindri et al. [63] prepared IL coatings of dicationic imidazolium containing different amino acid species as anionic entities with an objective to evaluate the impact of each amino acid species on the

Ionic liquids and Poly(ionic liquid)s Based Coatings   47

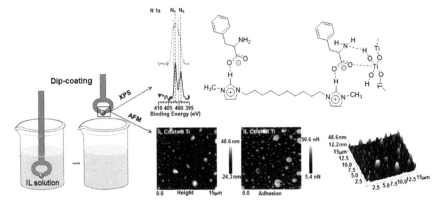

Figure 2.6 Ionic liquid coatings for titanium surfaces. Reproduced from Reference 60 with permission from American Chemical Society.

surface properties and passivation of titanium. Resistance against corrosion and wear was observed for the coating systems containing water repelling anionic species like phenylalanine and methionine in various solutions such as synthetic saliva, phosphate-buffered saline and their combination.

Gindri *et al.* [64] also reported the surface functionalization of TiO_2 nanoparticles using IL coatings based on dicationic imidazolium containing different anionic moieties. An enhancement in the IL coating thickness on TiO_2 surface was observed with an increase in IL hydrophobicity. The existence of appreciable adhesion of IL coating on TiO_2 surface was also confirmed. Figure 2.7 shows the chemical structures of the ionic liquids used in the study.

2.3.2 Ionic Liquids Based Coatings for Copper

For the effective passivation of copper, many research studies have employed IL based coatings. In one such study, Scendo *et al.* [65] discussed the usefulness of two IL coating systems based on 1-butyl-3-methylimidazolium chloride (BMIMCl) and 1-butyl-3-methylimidazolium bromide (BMIMBr) for the passivation of copper in acidic environments, especially in chloride medium. Both IL based coating systems exhibited anti-corrosion performance, which was observed to strengthen further with increase in IL concentration. Pisarova *et al.* [66] examined the effect of three different ILs butyl-trimethyl-ammonium bis(trifluoromethylsulfonyl)imide, cholin bis(trifluoromethylsulfonyl)imide and cholin methanesulfonate for passivation of copper alloy. Among the three ILs, cholin bis(trifluoromethylsulfonyl)imide based coating system demonstrated superior anti-corrosion potential.

Compound	IL structure
IL-1	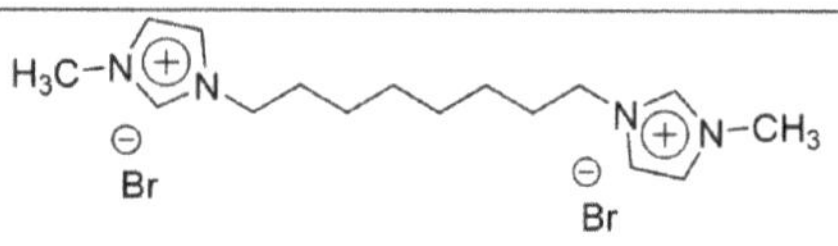
IL-2	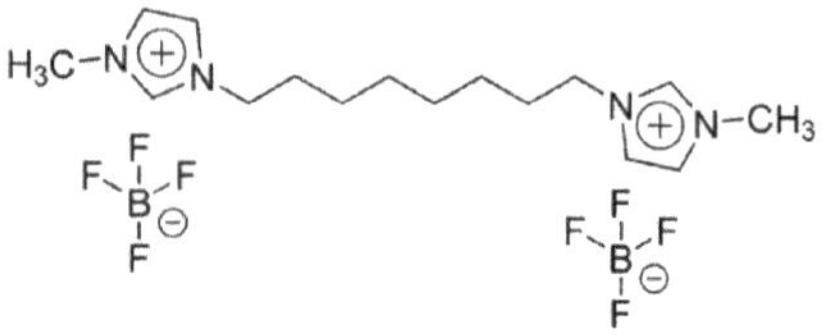
IL-3	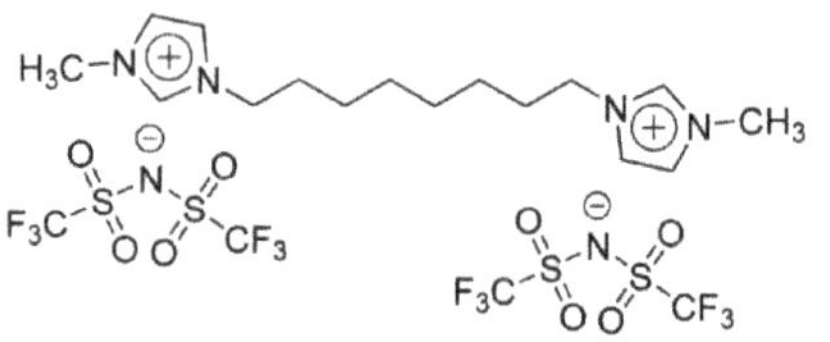

Figure 2.7 Chemical structures of the ionic liquids. Reproduced from Reference 64 with permission from American Chemical Society.

Espinosa et al. [67] employed different protic and aprotic ILs based coatings for the purpose of lubrication and passivation of oxygen-free highly thermally conductive copper surfaces. The protic ILs included triprotic di[(2-hydroxyethyl)ammonium] succinate and diprotic di[bis-(2-hydroxyethyl)ammonium] adipate. On the other hand, the aprotic ILs were 1-ethyl-3-methylimidazolium phosphonate ([EMIM]EtPO$_3$H), 1-ethyl-3-methylimidazolium octylsulfate ([EMIM]C$_8$H$_{17}$SO$_4$), 1-hexyl-3-methylimidazolium tetrafluoroborate ([HMIM]BF$_4$) and 1-hexyl-3-methylimidazolium hexafluorophosphate ([HMIM]PF$_6$). In the case of protic ILs based coatings, diprotic ammonium adipate based IL presented superior anti-corrosion performance together with appreciable lubrication effect and abrasion resistance owing to the creation of uniformly adsorbed film. For aprotic ILs based coatings, 1-hexyl-3-methylimidazolium hexafluorophosphate IL based coating displayed superior corrosion inhibition with lubrication effect and minimal abrasion due to the generation of fault-free film owing to adhesion with copper surface.

2.3.3 Ionic Liquids Based Coatings for Steel

A number of IL based coatings have been introduced for the passivation and enrichment of anti-corrosion property of various steel substrates. Yousefi *et al.* [68] studied the effect of coatings of various imidazolium based IL species, such as 1-ethyl-3-methylimidazolium chloride, 1-butyl-3-methylimidazolium chloride, 1-butyl-3-methylimidazolium hexafluorophosphate, 1-butyl-3-methylimidazolium tetrafluoroborate, 1-butyl-3-methylimidazolium bromide and 1-hexyl-3-methylimidazolium chloride, in combination with sodium dodecyl sulfate on the passivation of mild steel. IL coating system based on 1-hexyl-3-methylimidazolium chloride displayed the highest anti-corrosion potential on the mild steel surface in HCl medium due to the generation of crosslinked network of chloride and imidazolium. Also, the length of imidazolium chain was observed to exhibit influence on the passivation of mild steel.

Zheng *et al.* [69] reported the use of 1-octyl-3-methylimidazolium bromide and 1-allyl-3-octylimidazolium bromide ILs coatings for mild steel. Effectiveness of both imidazolium based coating systems for passivation and corrosion inhibition of mild steel surfaces in sulfuric acid containing medium was observed. Among the two, 1-allyl-3-octylimidazolium bromide exhibited greater extent of anti-corrosion activity owing to the influence of allyl group with substantial electron donating nature. In order to passivate mild steel surfaces, Ortaboy [70] reported the use of various phosphonium based ILs in combination with polyaniline by means of electrodeposition technique. The ILs based composite coatings were observed to be beneficial for improving the anti-corrosion properties of mild steel.

Tuken *et al.* [71] made use of IL coating system based on 1-ethyl-3-methylimidazolium dicyanamide for the passivation of mild steel surface. Tribological analysis revealed prominent potential of the IL for surface corrosion protection of mild steel in sulfuric acid solution. The IL also exhibited suitability as corrosion inhibitor in polypyrrole coating for achieving mild steel passivation. Shukla *et al.* [72] carried out a comparative study on the passivation effect of various aprotic IL coatings including [C$_4$mim][NO$_3$], [C$_4$mim][BF$_4$] and [C$_4$C$_1$mim][BF$_4$] on the surface of mild steel. The existence of optimal anti-corrosion activity was noted for IL coating of [C$_4$mim][NO$_3$] as compared to others due to the formation of uniform surface film. Ibrahim et al. [73] also discussed the usefulness of ILs based on imidazolium and pyridinium as effective corrosion protection coating systems for carbon steel owing to their adhesion and corresponding obstruction of reactive sites responsible for corrosion. Also, an increase in anti-corrosion behavior was observed with increase in IL concentration. In another study, Zhou et al. [74] achieved outstanding passivation of carbon steel through the application of IL coating based on [BMIM]BF$_4$. Kowsari et al. [75] also reported superior anti-corrosion activity for coating system comprising task specific imidazolium based IL on low carbon steel in hydrochloric acid containing corrosion medium.

2.3.4 Ionic Liquids Based Coatings for Aluminum

Owing to outstanding performance along with environmental sustainability, coatings based on ILs have received immense research attention to strengthen the tribological features and corrosion resistance of aluminum based substrates. In a recent study, Jimenez et al. [76] deployed imidazolium phosphonate and phosphate ionic liquids for the surface modification of aluminum alloy through the development of surface layers by electrochemical means. The application of ILs enhanced the tribological performance, especially, the abrasion resistance of aluminum alloy through the formation of uniform surface layer. Qiao et al. [77] also reported the advantages of coating systems based on butylammonium and tetrabutylammonium ILs with dibutylphosphate anionic moieties for enriching the surface features of aluminum alloy in contact with steel. Both IL based coating systems exhibited potential as lubricating coatings with significant resistance against wear due to reduction in coefficient of friction.

In another study, Huang et al. [78] described the passivation and anti-corrosion performance of aluminum alloy by the application of IL coating based on trihexyl(tetradecyl)phosphonium bis(trifluoromethylsulfonyl)amide using electrochemical techniques. Also, the coating system based on trihexyl(tetradecyl)phosphonium bis(trifluoromethylsulfonyl)imide ([P$_{6,6,6,14}$][NTf$_2$]) ionic liquid generated superior anti-corrosive capability on the surface of aluminum alloy [79]. In another strategy, the authors also disclosed the effectiveness of trihexyl(tetradecyl) phosphonium diphenylphosphate ([P 6, 6, 6, 14][dpp]) IL for the surface corrosion inhibition of aluminum alloy by means of passive layer generation.

In a recent study, Tateishi et al. [81] reported the usefulness of various anionic moieties in 1-butyl-3-methylimidazolium trifluoromethylacetate (BMIm-TFA), 1-butyl-3-methylimidazolium bis(trifluoromethylsulfonyl)amide (BMIm-TFSA) and 1-ethyl-3-methylimidazolium lactate (EMIm-LAC) for the passivation of aluminum surfaces. All ILs on the aluminum surface presented better anti-corrosion characteristics due to the generation of defect-free layers comprising of oxides and fluorides of aluminum. In another study, the authors revealed the effectiveness of three different ILs 1-butyl-3-methylimidazolium benzoate (BMIm-BEN), 1-butyl-3-methylimidazolium mandelate (BMIm-MAN) and 1-ethyl-3-methylimidazolium acetate (EMIm-ACE) for the formation of passive layer of aluminum oxide on aluminum surface. To enrich the anti-corrosive characteristics of aluminum based current collector of lithium-ion battery, Cho et al. [83] described passive layer formation through the application of bis(fluorosulfonyl)imide-based ionic liquid.

2.3.5 Ionic Liquids Based Coatings for Magnesium Alloy

Similar to other systems, many research studies have reported the application of ILs based coatings for the surface protection of magnesium alloys in order to enhance their lifespan. Espinosa et al. [84] reported the practicability of coatings based on ILs 1,3-dimethylimidazolium methylphosphonate, 1-ethyl-3-methylimidazolium methylphosphonate and 1-ethyl-3-methylimidazolium ethylphosphonate for the protection and wear resistance of magnesium alloy. Among the three ILs, coating system based on ethylphosphonate constituting IL exhibited maximum scratch resistance and protection capability owing to the formation of desirable thin layer coating with strong interaction with the magnesium alloy surface. Furthermore, the coatings generated at ambient temperature exhibited better performance as compared to coatings generated at 50 °C due to the existence of higher degree of adhesion with magnesium alloy surface. In another restudy, Jimenez et al. [85] utilized coating system based on room temperature IL 1-ethyl-3-methylimidazolium ethylphosphonate for the surface modification of magnesium alloy. The resultant coating demonstrated improved wear resistance (Figure 2.8) as well as reduced coefficient of friction due to the generation of uniform layer constituting phosphates of magnesium and aluminum.

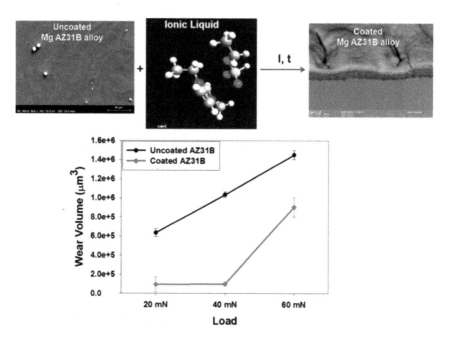

Figure 2.8 Representation of the surface coating of phosphonate ionic liquid for the enhancement of tribological performance. Reproduced from Reference 85 with permission from American Chemical Society.

Elsentriecy *et al.* [86] reported the application of IL coating based on aprotic ammonium phosphate for the effective passivation of magnesium alloy in sodium chloride containing corrosion medium. Coatings applied at 300 °C exhibited superior anti-corrosion performance as compared to coatings applied at ambient temperature owing to the formation of surface layer constituting phosphates and oxides of magnesium alloy through the reaction of IL thermal breakdown products on the metal surface. Also, the application of surface pretreatment was observed to be beneficial in enhancing the anti-corrosion activity of IL coatings. In another study, the authors observed almost similar passivation behavior for IL coating based on protic ammonium phosphate on magnesium alloy under same conditions [87]. In addition to oxides and phosphates of magnesium alloy in the conversion coating, the presence of carbonaceous compounds was also observed. Gu et al. [88] also described the usage of IL consisting of choline chloride and urea for the passivation of magnesium alloy. The application of IL on the alloy surface was useful to enrich the anti-corrosive nature through the generation of a fault-free surface layer containing hydride and carbonate of magnesium.

2.3.6 Other Functional Coatings Based on ILs

Gusain et al. [89] reported the development of environmentally friendly lubricant coating system based on imidazolium/ammonium-bis(salicylato)borate ionic liquids. The resultant coating system exhibited outstanding anti-wear characteristics owing to the decrease in coefficient of friction, along with superior corrosion resistance. In another study, Zhou et al. [90] reported the fabrication of polyalphaolefin (PAO) coatings possessing excellent resistance against wear through the introduction of a series of ILs. Among these ILs, organophosphate based IL system demonstrated optimal wear resistance and superior surface protection compared to coatings based on both sulfonate and carboxylate ILs due to the successful generation of phosphorus comprising surface layer (Figure 2.9). Gusain et al. [91] further reported the generation of green coating system based on fatty acid containing alkylammonium ILs in

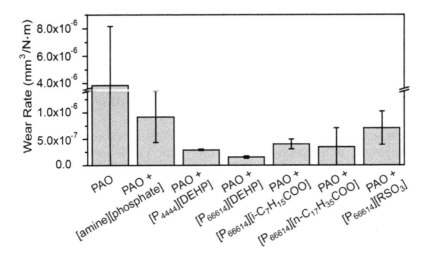

Figure 2.9 Wear rate of PAO and PAO-IL blends. Reproduced from Reference 90 with permission from American Chemical Society.

order to facilitate the formation of lubricant surface possessing outstanding anti-wear performance. Among different fatty acid anionic moieties based ILs, oleate based system exhibited superior tribological performance.

Landauer et al. [92] designed a series of coating systems comprising of phosphonium-organophosphate or ammonium-organophosphate IL individually and in combination with zinc dialkyldithiophosphate for developing high performance tribolayers on metallic surfaces. The resultant self-healing coatings exhibited superior anti-wear and low friction characteristics together with appreciable thermo-mechanical properties. In another study, Qu et al. [93] also reported the utilization of a coating system comprising of phosphonium-organophosphate ionic liquid and zinc dialkyldithiophosphate for the generation of high performance tribolayers on the non-metallic surfaces.

Fashu et al. [94] reported the development of a coating based on choline-chloride IL for the effective passivation of zinc-nickel-phosphorus alloy by means of electrodeposition technique. The resultant IL coating provided effective anti-corrosion performance on alloy surface. In another study, Siddiqui et al. [62] utilized dicationic imidazolium based IL coatings for the modification of cobalt-chromium-molybdenum alloy surface. The implementation of IL coatings was beneficial to strengthen the resistance against corrosion and abrasion. To enhance the performance of membranes based on polyacrylonitrile for the effective separation of gas molecules, Grunauer et al. [95] also reported the incorporation of 1-ethyl-3-methylimidazolium tetracyanoborate IL using dip coating technique.

2.4 Ionic Liquids Based Electrodeposited Al Coatings

Electrodeposition is one of the commonly used techniques for the development of protective Al coating over corrosion prone substrates by means of electrochemical reactions in an electrolytic solution to completely elude or diminish the corrosion tendency. Based on the nature of electrolytic solutions, the methods followed for the electrodeposition of Al coatings are categorized as electrochemical aluminization (ECA) and electrodeposition from ionic liquids (ECX). The former uses toluene and the latter ILs as electrolytes for the generation of anti-corrosive Al coatings. The coatings based on ECX method generally exhibit uniform morphology [96,97]. Together with non-volatility and high performance of ILs, Al electrodeposition by means of ECX method has been applied for the passivation of various substrates such as steel, copper, gold, magnesium alloy, NdFeB magnets, etc.

2.4.1 Coatings for Steel

Wulf *et al.* [97] reported the passivation of steel with the aid of electrodeposited Al from the fusion of $AlCl_3$ and 1-ethyl-3-methyl-imidazolium chloride ([Emim]Cl). The application of electrodeposited Al coating on the steel surface resulted in anti-corrosion character, which was analyzed using immersion in moving corrosive medium containing Pb and Li under high temperature conditions. The observed behavior was attributed to the existence of uniform morphology of ECX electroplated Al covering on steel [98].

Xue *et al.* [99] also reported the generation of aluminide coating on the interior of steel tubes by means of thermal treatment of Al electrodeposits from IL composed of 1-ethyl-3-methyl-imidazolium chloride and $AlCl_3$. Formation of continuous layer with robust adhesive strength on the steel surface was observed. In another study, El Abedin *et al.* [100] reported the utilization of above mentioned IL coating system for the surface modification of steel screw substrates. Improvement in adhesion of Al electrodeposits after surface pretreatment of steel screws was observed, which directly resulted in the enhancement of the anti-corrosion performance.

Bakkar *et al.* [101] described the surface modification of steel containing low amount of carbon using electrodeposited Al coating containing both nano- and micro-crystallites yielded from a mixture of 1-ethyl-3-methylimidazolium chloride ionic liquid and $AlCl_3$. The resultant coating system exhibited superior anti-corrosion characteristic in chloride based corrosive medium due to the creation of uniform surface layer with minimal defects. Nano-Al containing coatings were observed to have slightly higher corrosion resistance, as compared to micro-Al coatings.

To improve the anti-corrosive nature of mild steel surface, Xu *et al.* [102] reported the use of electroplated Al-Ti composite coating generated using a hybrid system composed of 1-buthyl-3-methylimidazolium chloride ionic liquid, $AlCl_3$ and $TiCl_4$. Electrochemical analysis confirmed the existence of higher anti-corrosion capability for Al-Ti composite coating compared to monolayer Al coating. In another study, Ispas *et al.* [103] also carried out the passivation of inferior carbon steel surface with the aid of electroplated Al monolayer as well as Al-Mi bilayer coatings generated using chloroaluminate ionic liquid, which resulted in the superior anticorrosion properties for both electrodeposited coatings in alkaline chloride medium.

2.4.2 Coatings for Copper

Li et al. [104] reported the facile formation of Al coatings on copper surface under ambient conditions by electroplating using a fusion of acetamide based eutectic IL and $AlCl_3$. In another study, Choudhary et al. [105] reported the modification of copper surface using electroplated Al coatings designed from a blend of [EMIm]Cl IL and $AlCl_3$. A study by Pulletikurthi et al. [106] detailed the surface modification of copper substrate using uniform thick Al coating generated from an IL based on 1-butylpyrrolidine and $AlCl_3$ by means of electroplating.

Tsuda et al. [107] reported the electrodeposited formation of Al-W layered coating on copper. To improve the performance of Al-W layered coating, the authors carried out the introduction of Mn into the coating composition. The creation of smooth surface morphology was fulfilled through the application of electroplating from a mixture of ethyl-3-methylimidazolium chloride IL and $AlCl_3$ accompanied by a combination of $MnCl_2$ and $K_3W_2Cl_9$. In another study, Gao et al. [108] used aluminum chloride ($AlCl_3$)/triethylamine hydrochloride (Et_3NHCl) ionic liquids for the development of effective electroplated Ni-Al bilayered coating on Cu surface. The resultant coating was observed to be continuous and dense with effective adhesion to the surface.

2.4.3 Coatings for Gold

Ismail [109] reported the utilization of IL (aryl-substituted imidazolium cation based IL) mixture of $AlCl_3$ for the generation of Al layers over gold surface. The exploitation of aryl imidazolium chloride IL resulted in the formation of surface film composed of Al nanocrystals. However. The adhesion strength of Al coatings electroplated from aryl imidazolium IL was observed to diminish as compared to alkyl imidazolium IL based Al coatings. In another study, the author also reported the generation of electroplated Al-Cu layered coating structure on gold surfaces through the use of 1-butyl-1-

methylpyrrolidinium bis (trifluoromethylsulfonyl) imide ionic liquid amalgam of $AlCl_3$. Suneesh et al. [111] described the formation of Al-Cu multilayered as well as Al monolayered coatings on gold surface with the aid of electroplating from an electrolytic solution based on triethylamine hydrochloride IL and $AlCl_3$. Improved corrosion resistance was observed for both coatings and Al-Cu layered coating structure exhibited slightly better anti-corrosion activity.

2.4.4 Coatings for Magnesium Alloy

For the passivation of magnesium alloy substrates, ECX electrodeposited Al coatings from 1-ethyl-3-methyl-imidazolium chloride IL have exhibited high potential. Onishi et al. [112] reported the modification of magnesium alloy using electrodeposited Al coating from an electrolytic solution based on 1-ethyl-3-methyl-imidazolium chloride IL and $AlCl_3$ and observed an increase in adhesion degree in response to enrichment in Al concentration in the coating system. In another study, Tang et al. [113] reported improvement in adhesion of electroplated Al coating generated from a mixture of 1-ethyl-3-methyl-imidazolium chloride and $AlCl_3$ on magnesium alloy through the application of surface pretreatment by zincate means. This resulted in enhanced anti-corrosion properties of Al coated magnesium alloy in alkaline chloride containing corrosive medium. Ueda et al. [114] reported appreciable improvement in anti-corrosion performance of magnesium alloy through the application of electrodeposited fault-free Al coating from 1-ethyl-3-methyl-imidazolium chloride IL and $AlCl_3$. Pan et al. [115] also reported the passivation of magnesium alloy substrate using electroplated Al-Zn layered coating with the aid of electrolyte system composed of 1-ethyl-3-methyl-imidazolium chloride IL, $AlCl_3$ and $ZnCl_2$. Enhancement in both anti-corrosion performance and hardness for thermally managed Al-Zn layered coatings was observed.

2.4.5 Coatings for Neodymium Iron Boron (NdFeB) Magnet

Jiang et al. [116] reported Al coating on the surface of NdFeB magnet using a mixture of 1-ethyl-3-methyl-imidazolium chloride IL and $AlCl_3$ by means of electroplating. In a related study, Xu et al. [117] also presented an enhancement in deposition amount as well as diminution in surface irregularity and grain size of these electroplated Al coatings through the increment in pulse rate.

Chen et al. [118] reported the use of a mixture of 1-ethyl-3-methyl-imidazolium chloride, $AlCl_3$ and $MnCl_2$ for the development of Al-Mn layered coating on NdFeB magnet by electroplating. Superior anti-corrosion activity in alkaline chloride containing corrosive medium along with approximately 3 folds decrease in corrosion current density was observed. In a related study, Ding et al. [119] also reported excellent passivation potential of these coatings along with enhanced micro-hardness and adhesion.

2.4.6 Coatings for Other Purposes

Li et al. [120] used a mixture of $AlCl_3$ and amide based IL for the formation of aluminum coating ov tungsten surface via electroplating technique. In another study, Li et al. [121] made use of $AlCl_3$/1-ethyl-3-methyl-imidazolium chloride IL blend for the generation of effective Al coating on the surface of tungsten electrode with an objective to upgrade the morphological features. The inclusion of $LaCl_3$ in the electrolyte formulation was beneficial to enhance the Al coating adhesion on tungsten surface with diminished porous nature.

Peng et al. [122] reported the surface modification of vanadium alloy through the application of aluminide coating, which was designed through the thermal treatment of ambient temperature electroplated Al coating obtained from a mixture of 1-ethyl-3-methyl-imidazolium chloride IL and $AlCl_3$. Fang et al. [123] utilized IL system composed of 4-propylpyridine and $AlCl_3$ for the generation of Al coating possessing superior corrosion resistance. The electrodeposited Al coating could be used for passivation of different materials to protect them from corrosion.

Cross and Schuh [124] made use of electrodeposition technique to design layered coating structure from a mixture of 1-ethyl-3-methyl-imidazolium chloride IL and $AlCl_3$. As a result, Al-Zn layered coating structure was obtained and its performance was further enhanced through the introduction of Zr at the time of electroplating. The resultant Al-Zn-Zr trilayered coating structure exhibited excellent anti-corrosion characteristic in alkaline chloride solution due to enhanced surface smoothness, diminished particle size and morphological co-existence of nano-crystallites as well as amorphous moieties. In another study, Tsuda et al. [125] reported the generation of Al-W layered coating structure through the use of aluminum chloride-1-ethyl-3-methylimidazolium chloride mixture in combination with $K_3W_2Cl_9$.

Table 2.2 summarizes various IL based electrodeposited Al coatings reported for the protection of different substrates.

Table 2.2 Summary of IL based electrodeposited Al coatings

Authors	Type of Al coating	Substrate
Wulf et al. [97]	Al coating	Steel
Xue et al. [99]	Aluminide coating from formed Al coating by thermal treatment	Steel tube interior
Bakkar et al. [101]	Al coating containing both nano and micro crystallites	Low carbon steel
Xu et al. [102]	Al-Ti composite coating	Mild steel
Ispas et al. [103]	Al monolayer as well as Al-Mi composite coatings	Low carbon steel
Li et al. [104]	Al coating	Copper
Choudhary et al. [105]	Al coating	Copper
Pulletikurthi et al. [106]	Flawless thick Al coating	Copper
Tsuda et al. [107]	Al-W-Mn composite coating	Copper
Gao et al. [108]	Flawless thick Ni-Al composite coating	Copper
Ismail [109]	Al coatings with nano- crystals	Gold
Ismail [110]	Al-Cu composite coating	Gold
Suneesh et al. [111]	Al monolayer as well as Al-Cu composite coatings	Gold
Onishi et al. [112], Tang et al. [113] and Ueda et al. [114]	Al coating	Magnesium alloy
Pan et al. [115]	Al-Zn composite coating	Magnesium alloy
Jiang et al. [116] and Xu et al. [117]	Al coating	NdFeB magnet
Chen et al. [118] and Ding et al. [119]	Al-Mn composite coating	NdFeB magnet
Li et al. [120]	Al coating	Tungsten
Li et al. [121]	Al coating	Tungsten
Peng et al. [122]	Al coating	Vanadium alloy

2.5 Conclusions

Coatings based on ILs and PILs have gained significant research attention owing to their unique performance profiles and environmentally friendliness. Specifically, ILs are preferred for diverse coatings due to their distinctive characteristics such as excellent flow resistance, ionic conductivity, substantial thermal, flame and chemical stability, good stability in electrochemical conditions, absence of volatility, high solubilizing capability, safe operating conditions, etc. PILs are upgraded class of ILs having the benefits of both ILs and macromolecules. In addition to usual features of ILs, PILs provide good mechanical and dimensional consistency, which are characteristics of macromolecules.

In this review, recent studies presenting the development of ILs and PILs based coatings for various applications have been summarized. Both ILs and PILs demonstrate suitability for a diverse range of applications such as protective coatings, sorbent coatings, lubricant coating, etc. These functional coatings are developed through the use of ILs and PILs, along with electrodeposition from ILs, etc., with the aid of various techniques such as dipping, spin coating and electrodeposition. The utilization of various ILs and PILs in the coating systems has been observed to enhance surface uniformity, corrosion resistance, wear resistance, lubrication and sorption characteristics of the substrates.

In summary, this study provides a current state of the art on the development of ILs and PILs based coatings. Owing to the immense enhancements achieved for these materials in the recent years, it is envisaged that the commercial applications of these advanced materials will exponentially rise in the near future.

References

1. Wilkes, J. S., and Zaworotko, M. J. (1992) Air and water stable 1-ethyl-3-methylimidazolium based ionic liquids. *Journal of the Chemical Society, Chemical Communications*, 965-967.
2. Wasserscheid, P., and Welton, T. (2008) *Ionic Liquids in Synthesis*, John Wiley & Sons, USA.
3. Caminiti, R., and Gontrani, L. (2014) *The Structure of Ionic Liquids*, Springer, Germany.
4. Benedetto, A., Bodo, E., Gontrani, L., Ballone, P., and Caminiti, R. (2014) Amino acid anions in organic ionic compounds. An ab initio

study of selected ion pairs. *The Journal of Physical Chemistry B,* **118**, 2471-2486.
5. Joseph, A., Zyla, G., Thomas, V. I., Nair, P. R., Padmanabhan, A., and Mathew, S. (2016) Paramagnetic ionic liquids for advanced applications: A review. *Journal of Molecular Liquids,* **218**, 319-331.
6. Plechkova, N. V., and Seddon, K. R. (2008) Applications of ionic liquids in the chemical industry. *Chemical Society Reviews,* **37**, 123-150.
7. Hough, W. L., and Rogers, R. D. (2007) Ionic liquids then and now: from solvents to materials to active pharmaceutical ingredients. *Bulletin of the Chemical Society of Japan,* **80**, 2262-2269.
8. Arce, A., Earle, M. J., Rodriguez, H., Seddon, K. R., and Soto, A. (2009) Bis {(trifluoromethyl) sulfonyl} amide ionic liquids as solvents for the extraction of aromatic hydrocarbons from their mixtures with alkanes: effect of the nature of the cation. *Green Chemistry,* **11**, 365-372.
9. Green, M. D., and Long, T. E. (2009) Designing imidazole-based ionic liquids and ionic liquid monomers for emerging technologies. *Polymer Reviews,* **49**, 291-314.
10. Armand, M., Endres, F., MacFarlane, D. R., Ohno, H., and Scrosati, B. (2009) Ionic-liquid materials for the electrochemical challenges of the future. *Nature materials,* **8**, 621-629.
11. Zhang, S., Zhang, Q., Zhang, Y., Chen, Z., Watanabe, M., and Deng, Y. (2016) Beyond solvents and electrolytes: ionic liquids-based advanced functional materials. *Progress in Materials Science,* **77**, 80-124.
12. Lee, J. H., Lee, J. S., Lee, J.-W., Hong, S. M., and Koo, C. M. (2013) Ion transport behavior in polymerized imidazolium ionic liquids incorporating flexible pendant groups. *European Polymer Journal,* **49**, 1017-1022.
13. Srour, H., Leocmach, M., Maffeis, V., Ghogia, A., Denis-Quanquin, S., Taberlet, N., Manneville, S., Andraud, C., Bucher, C., and Monnereau, C. (2016) Poly (ionic liquid) s with controlled architectures and their use in the making of ionogels with high conductivity and tunable rheological properties. *Polymer Chemistry,* **7**, 6608-6616.
14. Campetella, M., Bodo, E., Caminiti, R., Martino, A., D'Apuzzo, F., Lupi, S., and Gontrani, L. (2015) Interaction and dynamics of ionic liquids based on choline and amino acid anions. *The Journal of Chemical Physics,* **142**, 234502.
15. Zaitsau, D. H., Kabo, G. J., Strechan, A. A., Paulechka, Y. U., Tschersich, A., Verevkin, S. P., and Heintz, A. (2006) Experimental vapor pressures of 1-alkyl-3-methylimidazolium bis (trifluoromethylsulfonyl) imides and a correlation scheme for estimation of vaporization enthalpies of ionic liquids. *The Journal of Physical Chemistry*

A, **110**, 7303-7306.
16. Martins, M. A., Frizzo, C. P., Moreira, D. N., Zanatta, N., and Bonacorso, H. G. (2008) Ionic liquids in heterocyclic synthesis. *Chemical Reviews,* **108**, 2015-2050.
17. Frizzo, C. P., Gindri, I. M., Bender, C. R., Tier, A. Z., Villetti, M. A., Rodrigues, D. C., Machado, G., and Martins, M. A. (2015) Effect on aggregation behavior of long-chain spacers of dicationic imidazolium-based ionic liquids in aqueous solution. *Colloids and Surfaces A: Physicochemical and Engineering Aspects,* **468**, 285-294.
18. Shirota, H., Mandai, T., Fukazawa, H., and Kato, T. (2011) Comparison between dicationic and monocationic ionic liquids: liquid density, thermal properties, surface tension, and shear viscosity. *Journal of Chemical & Engineering Data,* **56**, 2453-2459.
19. Wang, J., Yao, H., Nie, Y., Zhang, X., and Li, J. (2012) Synthesis and characterization of the iron-containing magnetic ionic liquids. *Journal of Molecular Liquids,* **169**, 152-155.
20. Yuan, J., Mecerreyes, D., and Antonietti, M. (2013) Poly(ionic liquid)s: an update. *Progress in Polymer Science,* **38**, 1009-1036.
21. Ikeda, T., Moriyama, S., and Kim, J. (2016) Quaternary ammonium cation functionalized poly(Ionic Liquid)s with poly(ethylene oxide) main chains. *Macromolecular Chemistry and Physics,* **217**, 2551-2557.
22. Ye, Y.-S., Rick, J., and Hwang, B.-J. (2013) Ionic liquid polymer electrolytes. *Journal of Materials Chemistry A,* **1**, 2719-2743.
23. Eftekhari, A., and Saito, T. (2017) Synthesis and properties of polymerized ionic liquids. *European Polymer Journal,* **90**, 245-272.
24. Yuan, J., and Antonietti, M. (2011) Poly(ionic liquid)s: polymers expanding classical property profiles. *Polymer,* **52**, 1469-1482.
25. Mecerreyes, D. (2011) Polymeric ionic liquids: Broadening the properties and applications of polyelectrolytes. *Progress in Polymer Science,* **36**, 1629-1648.
26. Choi, U. H., Lee, M., Wang, S., Liu, W., Winey, K. I., Gibson, H. W., and Colby, R. H. (2012) Ionic conduction and dielectric response of poly (imidazolium acrylate) ionomers. *Macromolecules,* **45**, 3974-3985.
27. Chen, Q., Masser, H., Shiau, H.-S., Liang, S., Runt, J., Painter, P. C., and Colby, R. H. (2014) Linear viscoelasticity and Fourier transform infrared spectroscopy of polyether–ester–sulfonate copolymer ionomers. *Macromolecules,* **47**, 3635-3644.
28. Kuan, W.-F., Remy, R., Mackay, M. E., and Epps III, T. H. (2015) Controlled ionic conductivity via tapered block polymer electrolytes. *RSC Advances,* **5**, 12597-12604.
29. Nykaza, J. R., Ye, Y., and Elabd, Y. A. (2014) Polymerized ionic liquid diblock copolymers with long alkyl side-chain length. *Polymer,* **55**, 3360-3369.

30. Liang, S., O'Reilly, M. V., Choi, U. H., Shiau, H.-S., Bartels, J., Chen, Q., Runt, J., Winey, K. I., and Colby, R. H. (2014) High ion content siloxane phosphonium ionomers with very low T_g. *Macromolecules*, **47**, 4428-4437.
31. Xue, Z., He, D., and Xie, X. (2015) Poly (ethylene oxide)-based electrolytes for lithium-ion batteries. *Journal of Materials Chemistry A*, **3**, 19218-19253.
32. Mori, H., Kudo, E., Saito, Y., Onuma, A., and Morishima, M. (2010) RAFT polymerization of vinyl sulfonate esters for the controlled synthesis of poly (lithium vinyl sulfonate) and sulfonated block copolymers. *Macromolecules*, **43**, 7021-7032.
33. Williams, S. R., Borgerding, E. M., Layman, J. M., Wang, W., Winey, K. I., and Long, T. E. (2008) Synthesis and characterization of well-defined 12, 12-ammonium ionenes: Evaluating mechanical properties as a function of molecular weight. *Macromolecules*, **41**, 5216-5222.
34. Lee, M., Choi, U. H., Salas-de la Cruz, D., Mittal, A., Winey, K. I., Colby, R. H., and Gibson, H. W. (2011) Imidazolium polyesters: structure–property relationships in thermal behavior, ionic conductivity, and morphology. *Advanced Functional Materials*, **21**, 708-717.
35. Saito, J., Miyatake, K., and Watanabe, M. (2008) Synthesis and properties of polyimide ionomers containing 1 H-1, 2, 4-triazole groups. *Macromolecules*, **41**, 2415-2420.
36. Visentin, A. F., Alimena, S., and Panzer, M. J. (2014) Influence of Ionic liquid selection on the properties of poly(ethylene glycol)diacrylate-supported ionogels as solid electrolytes. *ChemElectroChem*, **1**, 718-721.
37. Li, H., Wang, C., Liao, X., Xie, M., and Sun, R. (2017) Hybrid triazolium and ammonium ions-contained hyperbranched polymer with enhanced ionic conductivity. *Polymer*, **112**, 297-305.
38. Obadia, M. M., Mudraboyina, B. P., Allaoua, I., Haddane, A., Montarnal, D., Serghei, A., and Drockenmuller, E. (2014) Accelerated solvent-and catalyst-free synthesis of 1, 2, 3-triazolium-based poly(Ionic liquid)s. *Macromolecular Rapid Communications*, **35**, 794-800.
39. Xu, W., Ledin, P. A., Shevchenko, V. V., and Tsukruk, V. V. (2015) Architecture, assembly, and emerging applications of branched functional polyelectrolytes and poly(ionic liquid)s. *ACS Applied Materials & Interfaces*, **7**, 12570-12596.
40. Taghavikish, M., Subianto, S., Dutta, N. K., de Campo, L., Mata, J. P., Rehm, C., and Choudhury, N. R. (2016) Polymeric ionic liquid nanoparticle emulsions as a corrosion inhibitor in anticorrosion coatings. *ACS Omega*, **1**, 29-40.
41. Taghavikish, M., Subianto, S., Dutta, N. K., and Choudhury, N. R. (2015) Facile fabrication of polymerizable ionic liquid based-gel

beads via thiol–ene chemistry. *ACS Applied Materials & Interfaces,* **7**, 17298-17306.

42. Taghavikish, M., Subianto, S., Dutta, N. K., and Roy Choudhury, N. (2016) Novel thiol-ene hybrid coating for metal protection. *Coatings,* **6**, 17.

43. Cagliero, C., Ho, T. D., Zhang, C., Bicchi, C., and Anderson, J. L. (2016) Determination of acrylamide in brewed coffee and coffee powder using polymeric ionic liquid-based sorbent coatings in solid-phase microextraction coupled to gas chromatography-mass spectrometry. *Journal of Chromatography A,* **1449**, 2-7.

44. Cagliero, C., Nan, H., Bicchi, C., and Anderson, J. L. (2016) Matrix-compatible sorbent coatings based on structurally-tuned polymeric ionic liquids for the determination of acrylamide in brewed coffee and coffee powder using solid-phase microextraction. *Journal of Chromatography A,* **1459**, 17-23.

45. Ho, T. D., Toledo, B. R., Hantao, L. W., and Anderson, J. L. (2014) Chemical immobilization of crosslinked polymeric ionic liquids on nitinol wires produces highly robust sorbent coatings for solid-phase microextraction. *Analytica Chimica Acta,* **843**, 18-26.

46. Pacheco-Fernandez, I., Najafi, A., Pino, V., Anderson, J. L., Ayala, J. H., and Afonso, A. M. (2016) Utilization of highly robust and selective crosslinked polymeric ionic liquid-based sorbent coatings in direct-immersion solid-phase microextraction and high-performance liquid chromatography for determining polar organic pollutants in waters. *Talanta,* **158**, 125-133.

47. Cordero-Vaca, M., Trujillo-Rodríguez, M. J., Zhang, C., Pino, V., Anderson, J. L., and Afonso, A. M. (2015) Automated direct-immersion solid-phase microextraction using crosslinked polymeric ionic liquid sorbent coatings for the determination of water pollutants by gas chromatography. *Analytical and Bioanalytical Chemistry,* **407**, 4615-4627.

48. Joshi, M. D., Ho, T. D., Cole, W. T., and Anderson, J. L. (2014) Determination of polychlorinated biphenyls in ocean water and bovine milk using crosslinked polymeric ionic liquid sorbent coatings by solid-phase microextraction. *Talanta,* **118**, 172-179.

49. Zhang, C., and Anderson, J. L. (2014) Polymeric ionic liquid bucky gels as sorbent coatings for solid-phase microextraction. *Journal of Chromatography A,* **1344**, 15-22.

50. Trujillo-Rodriguez, M. J., Yu, H., Cole, W. T., Ho, T. D., Pino, V., Anderson, J. L., and Afonso, A. M. (2014) Polymeric ionic liquid coatings versus commercial solid-phase microextraction coatings for the determination of volatile compounds in cheeses. *Talanta,* **121**, 153-162.

51. Yu, H., Merib, J., and Anderson, J. L. (2016) Crosslinked polymeric ionic liquids as solid-phase microextraction sorbent coatings for

high performance liquid chromatography. *Journal of Chromatography A,* **1438**, 10-21.
52. Jia, J., Liang, X., Wang, L., Guo, Y., Liu, X., and Jiang, S. (2013) Nanoporous array anodic titanium-supported co-polymeric ionic liquids as high performance solid-phase microextraction sorbents for hydrogen bonding compounds. *Journal of Chromatography A,* **1320**, 1-9.
53. Chen, C., Liang, X., Wang, J., Zou, Y., Hu, H., Cai, Q., and Yao, S. (2014) Development of a polymeric ionic liquid coating for direct-immersion solid-phase microextraction using polyhedral oligomeric silsesquioxane as cross-linker. *Journal of Chromatography A,* **1348**, 80-86.
54. Roessler, A., and Schottenberger, H. (2014) Antistatic coatings for wood-floorings by imidazolium salt-based ionic liquids. *Progress in Organic Coatings,* **77**, 579-582.
55. Iwata, T., Tsurumaki, A., Tajima, S., and Ohno, H. (2014) Fixation of ionic liquids into polyether-based polyurethane films to maintain long-term antistatic properties. *Polymer,* **55**, 2501-2504.
56. Isik, M., Gracia, R., Kollnus, L. C., Tome, L. C., Marrucho, I. M., and Mecerreyes, D. (2013) Cholinium-based poly(ionic liquid)s: synthesis, characterization, and application as biocompatible ion gels and cellulose coatings. *ACS Macro Letters,* **2**, 975-979.
57. Isik, M., Gracia, R., Kollnus, L. C., Tome, L. C., Marrucho, I. M., and Mecerreyes, D. (2014) Cholinium lactate methacrylate: Ionic liquid monomer for cellulose composites and biocompatible ion gels. *Macromolecular Symposia,* **342**(1), 21-24.
58. Tokuda, M., Sanada, T., Shindo, T., Suzuki, T., and Minami, H. (2014) Preparation of submicrometer-sized quaternary ammonium-based poly(ionic liquid) particles via emulsion polymerization and switchable responsiveness of emulsion film. *Langmuir,* **30**, 3406-3412.
59. Gindri, I. M., Palmer, K. L., Siddiqui, D. A., Aghyarian, S., Frizzo, C. P., Martins, M. A., and Rodrigues, D. C. (2016) Evaluation of mammalian and bacterial cell activity on titanium surface coated with dicationic imidazolium-based ionic liquids. *RSC Advances,* **6**, 36475-36483.
60. Gindri, I. M., Siddiqui, D. A., Frizzo, C. P., Martins, M. A., and Rodrigues, D. C. (2015) Ionic liquid coatings for titanium surfaces: effect of IL structure on coating profile. *ACS Applied Materials & Interfaces,* **7**, 27421-27431.
61. Frizzo, C. P., Wust, K., Tier, A. Z., Beck, T. S., Rodrigues, L. V., Vaucher, R. A., Bolzan, L. P., Terra, S., Soares, F., and Martins, M. A. (2016) Novel ibuprofenate-and docusate-based ionic liquids: emergence of antimicrobial activity. *RSC Advances,* **6**, 100476-100486.

62. Siddiqui, D. A., Gindri, I. M., and Rodrigues, D. C. (2016) Corrosion and wear performance of titanium and cobalt chromium molybdenum alloys coated with dicationic imidazolium-based ionic liquids. *Journal of Bio-and Tribo-Corrosion*, **2**, 27.
63. Gindri, I. M., Siddiqui, D. A., Frizzo, C. P., Martins, M. A., and Rodrigues, D. C. (2016) Improvement of tribological and anticorrosive performance of titanium surfaces coated with dicationic imidazolium-based ionic liquids. *RSC Advances*, **6**, 78795-78802.
64. Gindri, I. M., Frizzo, C. P., Bender, C. R., Tier, A. Z., Martins, M. A., Villetti, M. A., Machado, G., Rodriguez, L. C., and Rodrigues, D. C. (2014) Preparation of TiO_2 nanoparticles coated with ionic liquids: a supramolecular approach. *ACS Applied Materials & Interfaces*, **6**, 11536-11543.
65. Scendo, M., and Uznanska, J. (2011) The effect of ionic liquids on the corrosion inhibition of copper in acidic chloride solutions. *International Journal of Corrosion*, Article ID 718626.
66. Pisarova, L., Gabler, C., Dorr, N., Pittenauer, E., and Allmaier, G. (2012) Thermo-oxidative stability and corrosion properties of ammonium based ionic liquids. *Tribology International*, **46**, 73-83.
67. Espinosa, T., Sanes, J., Jimenez, A.-E., and Bermudez, M.-D. (2013) Surface interactions, corrosion processes and lubricating performance of protic and aprotic ionic liquids with OFHC copper. *Applied Surface Science*, **273**, 578-597.
68. Yousefi, A., Javadian, S., Dalir, N., Kakemam, J., and Akbari, J. (2015) Imidazolium-based ionic liquids as modulators of corrosion inhibition of SDS on mild steel in hydrochloric acid solutions: experimental and theoretical studies. *RSC Advances*, **5**, 11697-11713.
69. Zheng, X., Zhang, S., Li, W., Gong, M., and Yin, L. (2015) Experimental and theoretical studies of two imidazolium-based ionic liquids as inhibitors for mild steel in sulfuric acid solution. *Corrosion Science*, **95**, 168-179.
70. Ortaboy, S. (2016) Electropolymerization of aniline in phosphonium-based ionic liquids and their application as protective films against corrosion. *Journal of Applied Polymer Science*, **133**(38), doi: 10.1002/app.43923.
71. Tuken, T., Demir, F., Kicir, N., Sigircik, G., and Erbil, M. (2012) Inhibition effect of 1-ethyl-3-methylimidazolium dicyanamide against steel corrosion. *Corrosion Science*, **59**, 110-118.
72. Shukla, S. K., Murulana, L. C., and Ebenso, E. E. (2011) Inhibitive effect of imidazolium based aprotic ionic liquids on mild steel corrosion in hydrochloric acid medium. *International Journal of Electrochemical Science*, **6**, 4286-4295.
73. Ibrahim, M. A., Messali, M., Moussa, Z., Alzahrani, A. Y., Alamry, S. N., and Hammouti, B. (2011) Corrosion inhibition of carbon steel by imidazolium and pyridinium cations ionic liquids in acidic envi-

ronment. *Portugaliae Electrochimica Acta,* **29**, 375-389.
74. Zhou, X., Yang, H., and Wang, F. (2011) [BMIM]BF$_4$ ionic liquids as effective inhibitor for carbon steel in alkaline chloride solution. *Electrochimica Acta,* **56**, 4268-4275.
75. Kowsari, E., Payami, M., Amini, R., Ramezanzadeh, B., and Javanbakht, M. (2014) Task-specific ionic liquid as a new green inhibitor of mild steel corrosion. *Applied Surface Science,* **289**, 478-486.
76. Jimenez, A.-E., Arias-Pardilla, J., Martinez-Nicolas, G., and Bermudez, M.-D. (2016) Electrochemical treatment of aluminium alloy 7075 in aqueous solutions of imidazolium phosphonate and phosphate ionic liquids and scratch resistance of the resultant materials. *Tribology International,* **113**, 65-75.
77. Qiao, D., Wang, H., and Feng, D. (2014) Tribological performance and mechanism of phosphate ionic liquids as additives in three base oils for steel-on-aluminum contact. *Tribology Letters,* **55**, 517-531.
78. Huang, P., Howlett, P. C., and Forsyth, M. (2014) Electrochemical etching of AA5083 aluminium alloy in trihexyl(tetradecyl)phosphonium bis(trifluoromethylsulfonyl)amide ionic liquid. *Corrosion Science,* **80**, 120-127.
79. Huang, P., Howlett, P., MacFarlane, D., and Forsyth, M. (2012) Passivation of AA5083 Aluminium Alloy by Anodic Pre-Treatments in Ionic Liquids. *ECS Meeting Abstracts,* 2101. Online: http://ma.ecsdl.org/content/MA2012-02/20/2101.abstract?cited-by=yes&legid=ecsmtgabs;MA2012-02/20/2101 (assessed 15th May 2018).
80. Huang, P., Somers, A., Howlett, P. C., and Forsyth, M. (2016) Film formation in trihexyl(tetradecyl) phosphonium diphenylphosphate ([P 6, 6, 6, 14][dpp]) ionic liquid on AA5083 aluminium alloy. *Surface and Coatings Technology,* **303**, 385-395.
81. Tateishi, K., Akiko, W., Ogino, H., Ohishi, T., and Murakami, M. (2012) Formation of Al$_2$O$_3$ film and AlF$_3$ containing Al$_2$O$_3$ film by an anodic polarization of aluminum in ionic liquids. *Electrochemistry,* **80**, 556-560.
82. Tateishi, K., Ogino, H., Akiko, W., Ohishi, T., Murakami, M., Hidetaka, A., and Sachiko, O. (2013) Anodization behavior of aluminum in ionic liquids with a small amount of water. *Electrochemistry,* **81**, 440-447.
83. Cho, E., Mun, J., Chae, O. B., Kwon, O. M., Kim, H.-T., Ryu, J. H., Kim, Y. G., and Oh, S. M. (2012) Corrosion/passivation of aluminum current collector in bis(fluorosulfonyl)imide-based ionic liquid for lithium-ion batteries. *Electrochemistry Communications,* **22**, 1-3.
84. Espinosa, T., Sanes, J., and Bermudez, M.-D. (2015) Halogen-free phosphonate ionic liquids as precursors of abrasion resistant surf-

ace layers on AZ31B magnesium alloy. *Coatings,* **5**(1), 39-53.
85. Jimenez, A. E., Rossi, A., Fantauzzi, M., Espinosa, T., Arias-Pardilla, J., Martinez-Nicolas, G., and Bermudez, M.a.-D. (2015) Surface coating from phosphonate ionic liquid electrolyte for the enhancement of the tribological performance of magnesium alloy. *ACS Applied Materials & Interfaces,* **7**, 10337-10347.
86. Elsentriecy, H. H., Luo, H., Meyer, H. M., Grado, L. L., and Qu, J. (2014) Effects of pretreatment and process temperature of a conversion coating produced by an aprotic ammonium-phosphate ionic liquid on magnesium corrosion protection. *Electrochimica Acta,* **123**, 58-65.
87. Elsentriecy, H. H., Qu, J., Luo, H., Meyer, H. M., Ma, C., and Chi, M. (2014) Improving corrosion resistance of AZ31B magnesium alloy via a conversion coating produced by a protic ammonium-phosphate ionic liquid. *Thin Solid Films,* **568**, 44-51.
88. Gu, C., Yan, W., Zhang, J., and Tu, J. (2016) Corrosion resistance of AZ31B magnesium alloy with a conversion coating produced from a choline chloride—Urea based deep eutectic solvent. *Corrosion Science,* **106**, 108-116.
89. Gusain, R., Singh, R., Sivakumar, K., and Khatri, O. P. (2014) Halogen-free imidazolium/ammonium-bis (salicylato) borate ionic liquids as high performance lubricant additives. *RSC Advances,* **4**, 1293-1301.
90. Zhou, Y., Dyck, J., Graham, T. W., Luo, H., Leonard, D. N., and Qu, J. (2014) Ionic liquids composed of phosphonium cations and organophosphate, carboxylate, and sulfonate anions as lubricant antiwear additives. *Langmuir,* **30**, 13301-13311.
91. Gusain, R., and Khatri, O. P. (2016) Fatty acid ionic liquids as environmentally friendly lubricants for low friction and wear. *RSC Advances,* **6**, 3462-3469.
92. Landauer, A. K., Barnhill, W. C., and Qu, J. (2016) Correlating mechanical properties and anti-wear performance of tribofilms formed by ionic liquids, ZDDP and their combinations. *Wear,* **354**, 78-82.
93. Qu, J., Meyer, H. M., Cai, Z.-B., Ma, C., and Luo, H. (2015) Characterization of ZDDP and ionic liquid tribofilms on non-metallic coatings providing insights of tribofilm formation mechanisms. *Wear,* **332**, 1273-1285.
94. Fashu, S., Gu, C., Wang, X., and Tu, J. (2014) Structure, composition and corrosion resistance of Zn-Ni-P alloys electrodeposited from an ionic liquid based on choline chloride. *Journal of the Electrochemical Society,* **161**, D3011-D3017.
95. Grunauer, J., Filiz, V., Shishatskiy, S., Abetz, C., and Abetz, V. (2016) Scalable application of thin film coating techniques for supported liquid membranes for gas separation made from ionic liquids. *Jou-*

rnal of Membrane Science, **518**, 178-191.
96. Wulf, S.-E., Krauss, W., and Konys, J. (2014) Comparison of coating processes in the development of aluminum-based barriers for blanket applications. *Fusion Engineering and Design,* **89**, 2368-2372.
97. Wulf, S.-E., Krauss, W., and Konys, J. (2016) Corrosion resistance of Al-based coatings in flowing Pb–15.7 Li produced by aluminum electrodeposition from ionic liquids. *Nuclear Materials and Energy,* **9**, 519-523.
98. Krauss, W., Konys, J., and Wulf, S.-E. (2014) Corrosion barriers processed by Al electroplating and their resistance against flowing Pb–15.7 Li. *Journal of Nuclear Materials,* **455**, 522-526.
99. Xue, D., Chen, Y., Ling, G., Liu, K., Chen, C. a., and Zhang, G. (2015) Preparation of aluminide coatings on the inner surface of tubes by heat treatment of Al coatings electrodeposited from an ionic liquid. *Fusion Engineering and Design,* **101**, 128-133.
100. El Abedin, S. Z., and Endres, F. (2013) Challenges in the electrochemical coating of high-strength steel screws by aluminum in an acidic ionic liquid composed of 1-ethyl-3-methylimidazolium chloride and $AlCl_3$, *Journal of Solid State Electrochemistry,* **17**, 1127-1132.
101. Bakkar, A., and Neubert, V. (2013) Electrodeposition and corrosion characterisation of micro- and nano-crystalline aluminium from $AlCl_3$/1-ethyl-3-methylimidazolium chloride ionic liquid. *Electrochimica Acta,* **103**, 211-218.
102. Xu, C., Hua, Y., Zhang, Q., Li, J., Lei, Z., and Lu, D. (2017) Electrodeposition of Al-Ti alloy on mild steel from $AlCl_3$-BMIC ionic liquid. *Journal of Solid State Electrochemistry,* **21**(5), 1349-1356.
103. Ispas, A., Vlaic, C. A., Camargo, M. K., and Bund, A. (2016) Electrochemical deposition of aluminum and aluminum-manganese alloys in ionic liquids. *ECS Transactions,* **75**, 657-665.
104. Li, M., Gao, B., Liu, C., Chen, W., Shi, Z., Hu, X., and Wang, Z. (2015) Electrodeposition of aluminum from $AlCl_3$/acetamide eutectic solvent. *Electrochimica Acta,* **180**, 811-814.
105. Choudhary, R., Rajak, S., Bidaye, A., Kain, V., and Hubli, R. (2013) Substrate and current density effects on electrodeposited aluminium coatings. *Surface Engineering,* **29**, 677-682.
106. Pulletikurthi, G., Bodecker, B., Borodin, A., Weidenfeller, B., and Endres, F. (2015) Electrodeposition of Al from a 1-butylpyrrolidine-$AlCl_3$ ionic liquid. *Progress in Natural Science: Materials International,* **25**, 603-611.
107. Tsuda, T., Ikeda, Y., Kuwabata, S., Stafford, G. R., and Hussey, C. L. (2014) Electrodeposition of Al-W-Mn alloy from Lewis acidic $AlCl_3$–1-ethyl-3-methylimidazolium chloride ionic liquid. *ECS Transactions,* **64**, 563-574.

108. Gao, L.-X., Wang, L.-N., Qi, T., and Yu, J. (2012) Preparation of Ni and Ni-Al Alloys from 2AlCl$_3$/Et3NHCl ionic liquid by electrodeposition. *Acta Physico-Chimica Sinica*, **28**, 111-120.
109. Ismail, A. S. (2015) Nano-sized aluminum coatings from aryl-substituted imidazolium cation based ionic liquid. *Egyptian Journal of Petroleum*, **25**(6), 525-530.
110. Ismail, A. S. (2017) Electrodeposition of aluminium–copper alloy from 1-butyl-1-methylpyrrolidinium bis (trifluoromethylsulfonyl) imide ionic liquid. *Egyptian Journal of Petroleum*, **26**, 61-65.
111. Suneesh, P. V., Satheesh Babu, T. G., and Ramachandran, T., (2013) Electrodeposition of aluminium and aluminium-copper alloys from a room temperature ionic liquid electrolyte containing aluminium chloride and triethylamine hydrochloride. *International Journal of Minerals, Metallurgy and Materials*, **20**, 909-916.
112. Onishi, M., Ueda, M., Matsushima, H., Habazaki, H., Washio, K., and Kato, A. (2016) Measurement of adhesion strength of Al electroplating film on AZ31, AZ61, and AZ91 substrates. *ECS Transactions*, **75**, 297-304.
113. Tang, J., and Azumi, K. (2012) Improvement of Al coating adhesive strength on the AZ91D magnesium alloy electrodeposited from ionic liquid. *Surface and Coatings Technology*, **208**, 1-6.
114. Ueda, M., Hariyama, S., and Ohtsuka, T. (2012) Al electroplating on the AZ121 Mg alloy in an EMIC–AlCl$_3$ ionic liquid containing ethylene glycol. *Journal of Solid State Electrochemistry*, **16**, 3423-3427.
115. Pan, S.-J., Tsai, W.-T., Kuo, J.-C., and Sun, I.-W. (2013) Material characteristics and corrosion performance of heat-treated Al-Zn coatings electrodeposited on AZ91D magnesium alloy from an ionic liquid. *Journal of the Electrochemical Society*, **160**, D320-D325.
116. Jiang, L., Jin, Z., Xu, F., Yu, Y., Wei, G., Ge, H., Zhang, Z., and Cao, C. (2016) Effect of potential on aluminium early-stage electrodeposition onto NdFeB magnet. *Surface Engineering*, **33**(5), 375-382.
117. Xu, F., Jiang, L., Wu, J., Shen, X., Huang, Q., Yu, Y., Cao, C., Wei, G., and Ge, H. (2017) Effects of pulsed-electrodeposition parameters on the property of aluminum film onto sintered NdFeB magnets. *International Journal of Electrochemical Science*, **12**, 517-528.
118. Chen, J., Xu, B., and Ling, G. (2012) Amorphous Al–Mn coating on NdFeB magnets: Electrodeposition from AlCl$_3$–EMIC–MnCl$_2$ ionic liquid and its corrosion behavior. *Materials Chemistry and Physics*, **134**, 1067-1071.
119. Ding, J., Xu, B., and Ling, G. (2014) Al–Mn coating electrodeposited from ionic liquid on NdFeB magnet with high hardness and corrosion resistance. *Applied Surface Science*, **305**, 309-313.
120. Li, M., Gao, B., Liu, C., Chen, W., Wang, Z., Shi, Z., and Hu, X. (2017) AlCl$_3$/amide ionic liquids for electrodeposition of aluminum. *Journal of Solid State Electrochemistry*, **21**(2), 469-476.

121. Li, B., Fan, C., Chen, Y., Lou, J., and Yan, L. (2011) Pulse current electrodeposition of Al from an $AlCl_3$-EMIC ionic liquid. *Electrochimica Acta,* **56**, 5478-5482.
122. Peng, X., Zhang, G., Yang, F., Xiang, X., Luo, L., Chen, C. a., and Wang, X. (2016) Fabrication and characterization of aluminide coating on V–5Cr–5Ti by electrodeposition and subsequent heat treating. *International Journal of Hydrogen Energy,* **41**, 8935-8945.
123. Fang, Y., Yoshii, K., Jiang, X., Sun, X.-G., Tsuda, T., Mehio, N., and Dai, S. (2015) An $AlCl_3$ based ionic liquid with a neutral substituted pyridine ligand for electrochemical deposition of aluminum. *Electrochimica Acta,* **160**, 82-88.
124. Cross, S. R., and Schuh, C. A. (2016) Ternary alloying additions and multilayering as strategies to enhance the galvanic protection ability of Al-Zn coatings electrodeposited from ionic liquid solution. *Electrochimica Acta,* **211**, 860-870.
125. Tsuda, T., Ikeda, Y., Arimura, T., Imanishi, A., Kuwabata, S., Hussey, C. L., and Stafford, G. (2013) Al-W alloy deposition from Lewis acidic room-temperature chloroaluminate ionic liquid. *ECS Transactions,* **50**, 239-250.

3

Carbon Based Coatings

3.1 Introduction

During the past few decades, carbon based coatings have gained immense importance for a wide range of industrial, commercial and domestic applications. Carbon exhibits wide variety of allotropes and structures because of its unique ability of hybridization which imparts it the status of a special element in the periodic table. Carbon-based materials have attracted immense interest, especially by invention of fullerenes, carbon nanotubes and graphene. By modifying bonding ratio of carbon from sp^3 to sp^2 and mixing or blending with other elements, the resultant unique physical and mechanical characteristics of carbon make it feasible for use in various demanding applications. For example, graphite coatings have been used since long as solid lubricants because of their low-friction coefficient. This is possible because of the sp^2 bonding present in them. In a similar way, the sp^3 bonding arrangement of "diamond like coatings" (DLC), which are high hardness coatings, has been used since early 1950s. Combining the properties of graphite coatings and DLC has also resulted in very efficient coatings with high order of mechanical and structural resistance [1].

In 1990s, the application of carbon based coatings found its way into more technological applications like fuel injection system in the diesel engines which provided safety against corrosion, wear and tear effects of the lubricant system. In the 21st century, carbon based coatings have become even more profound in their use. Almost every industry like pharmaceuticals, food packaging, and petroleum industry utilizes carbon based coatings for a wide spectrum of functional applications. Carbon coatings also find great utilization in mechanical industries where these are used to protect the surfaces from corrosion resulting from wear and tear of the metal surface; some of the examples include compressors and pumps, gears, shafts, etc. [2]. Besides this, carbon coatings are added on various metallic and non-metallic substrates which enhances the properties of the material; a

Ahmad Tabish and Vikas Mittal, Khalifa University of Science and Technology), Abu Dhabi, UAE
© 2021 Central West Publishing, Australia

common example being the application of carbon based coatings on lithium electrodes.

Carbon exists in various forms like carbon black, carbon nanomaterials, mesoporous carbon, amorphous carbon, microporous carbon, etc. The following sections review the various types of coatings generated by employing the active carbon component in various structural forms. The most recent findings in the synthesis, characterization and application of these carbon-based coatings are highlighted.

3.2 Mesoporous Carbon Coatings

Mesoporous materials have pores whose diameter ranges usually from 2 to 50 nm. Porous carbon materials are of interest in many applications because of their physio-chemical properties and high surface area. Due to their ordered mesostructures and high surface areas, these materials are widely used as catalysts, advanced electronic materials, separation media, etc. The most recent type of porous carbon materials are ordered mesoporous carbons (OMCs). The BET specific surface area of these materials is found to be around 2200 m^2/g. Moreover, these materials exhibit immense stability at high temperatures and have high resistance in acidic or basic media.

Recently, the mesoporous carbon based coatings have gained importance in various applications. The below sections summarize various research studies on the subject.

3.2.1 Fabrication

In a study by Mittal *et al.* [3], a mesoporous carbon nanocapsule (MCC)/polyvinylidene fluoride (PVDF) polymer composite based free-standing film was fabricated with superhydrophobic nature and multifunctionalities. Mesoporous carbon (MCC) capsules which were in micron range were synthesized by impregnating carbon precursor into the silica spheres which were later carbonized leading to the removal of silica from the latter. The PVDF films were synthesized by first dissolving 10 wt% of PVDF in N-methyl-2-pyrolidone (NMP) solvent. Later, 1H,1H,2H,2H-perfluorodecyltriethoxysilane (PFS) was added in the above solution. MCC was subsequently added to the polymer mixture by manually stirring. It was followed by dipping of metal substrate in the polymer solution and drying at 120 °C for around 2 hours.

Wan et al. [4] studied the adsorption application of ordered mesoporous carbon based coating over cordierite with honeycomb structure. Here, the mesoporous carbon was coated on cordierite as a carrier and the surfactant-templating method was used. Carbon sources used for the work were formaldehyde and some phenols. Moreover, copolymer F127 having triblock shape was utilized as a structure dictating agent. In a very similar fashion, another film of mesoporous carbon was synthesized with silicon wafer as substrate, instead of cordierite and the spin-coating method was employed instead of templating for the purpose of comparison of resulting films by the two methods.

Sol-gel spin-coating technique has also been utilized by Pang et al. [5] for synthesizing thin films based on mesoporous carbon. In this study, precursors used were tetraethyl orthosilicate (TEOS) and cheaply available sucrose. Thus, films based on the blend of sucrose and silica nanocomposites were synthesized by reacting TEOS in a solution having acidic sucrose, followed by spin coating the formed solution having both sucrose and silica. Later, the carbonization of the films was carried out which resulted in the formation of carbon/silica nanocomposite thin films. Finally, the removal of silica network was carried out using HF which resulted in the thin films based on mesoporous carbon.

In another study, coating-etching approach was used by Feng et al. [6]. In this study, similar mesoporous carbon based free-standing thin films were fabricated with arranged pore structures as well as varying diameters (Figure 3.1). For the synthesis of the film,

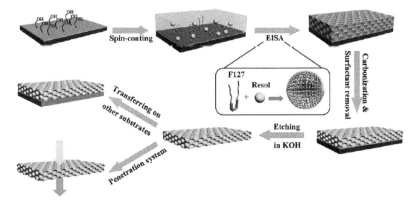

Figure 3.1 Schematic of the preparation of free-standing mesoporous carbon thin films. Reproduced from Reference 6 with permission from American Chemical Society.

preoxidized silicon wafer was used as a substrate, on which resol precursors/pluronic copolymer were coated. It resulted in the formation of polymeric mesostructures of highly ordered nature. Finally, carbonization was carried out at 600 °C which was followed by etching the layer of oxide formed between the silicon substrate and carbon film.

Another recent study on the generation of mesoporous carbon coatings has been reported by Mittal *et al.* [7]. In this study, mesoporous carbon nanocapsules (MCC) based super hydrophobic coatings have been fabricated which are multifunctional in nature. Just like previous studies, PVDF was used as a binder. A facile brush-on process was used on multiple substrates like cotton based fabric, metals and glass, etc. Solid mesoporous carbon coatings have also been synthesized in a similar manner in other research studies [8-10]. For instance, Figure 3.2 demonstrates metal-organic framework (IRMOF-1) after controlled pyrolysis for generating carbon-coated ZnO quantum dots [9].

Figure 3.2 Metal organic framework (IRMOF-1; top) and IRMOF-1 after controlled pyrolysis for generating carbon-coated ZnO QDs without agglomeration. Reproduced from Reference 9 with permission from American Chemical Society.

Another example of the mesoporous coating framework has been demonstrated by Ji et al. [11] in which centimeter-scale free standing thin films have been synthesized from 2D nanocrystal superlattices self-assembled at the solid– or liquid–air interface and are collectively called mesoporous graphene frameworks (MGFs). MGF thin films were obtained based on the transformation of 2D nanocrystal superlattices by ligand carbonization, nanocrystal etching, and framework graphitization [12].

3.2.2 Characterization and Testing

Mittal et al. [3] carried out the transmission electron microscopy (TEM) analysis of MCC with and without silica core for morphological and structural characterization. Moreover, field emission scanning electron microscopy (SEM) was also used for studying the free standing films (FSFs). SEM revealed homogeneously distributed membrane resembling a porous film with sub-micron sized pores varying from 0.5 to 3 um approx. in size on the top film surface. Higher magnification exhibited that mesoporous carbon nanocapsules were mostly around 200-300 nm in size. The SEM images of the edge view of the MCC/polymer based superhydrophobic FSF showed that the FSFs had a thickness of around 120 um.

Wan et al. [4] characterized the mesoporous carbon coating on cordierite by TEM, high-resolution scanning electron microscopy (HRSEM), and nitrogen sorption techniques which exhibited the presence of carbon coating pore arrays ordered hexagonally over cordierite. Also, ordered mesoporous carbon was used to coat the honeycomb monolith adsorbents [13] which exhibited adsorption capacities for chlorinated organic pollutants in water with 178 mg/g for p-chloroaniline and 200 mg/g for p-chlorophenol (with respect to the net carbon coating), along with adsorption of lower concentration contaminants, handling of high processing volumes and recyclability. Similarly, Pang et al. [5] used atomic force microscopy (AFM) and SEM images of mesoporous carbon films after the removal of silica template which indicated the appearance of smooth, homogeneous and continuous crack free thin films. Further, SEM indicated the film thickness to be 1 micron on an average. TEM study of mesoporous carbon films suggested that mesoporous carbon contained a non-ordered, but homogeneously sized mesoporous structure.

Mittal et al. [7] also characterized the morphology and structure of MCC based coatings by microscopy and optical profilometer. SEM exhibited the 3D structure of structured MCC with a narrow size dispersion (radius around 100-150 nm) revealing presence of pores formed within. Moreover, TEM demonstrated the presence of capsules of thin carbon shell before and after silica was etched out from the core. Fourier transform infrared (FTIR) spectroscopy was utilized to study the chemical nature of superhydrophobic coating. The spectra obtained after annealing at 100 °C for 10 hours exhibited peaks at 895, 1017, and 1138 cm^{-1}, which were attributed C–H bending modes, C–F stretching modes and definite variations of CF_2 and CF groups present in polyvinylidene fluoride.

Nitrogen sorption technique has been utilized by Feng *et al.* [6] to evaluate the mesoporosity of carbon films, which demonstrated a large BET surface area of 700 m^2/g and large homogeneous mesopores with average size of ~4.4 nm. The SEM analysis exhibited a homogenous and smooth free-standing film layer with no visible cracks. The film thickness was observed as 500 nm and was uniform in nature, as confirmed by both the surface profiler and SEM. The mesopore architecture of the synthesized free-standing mesoporous carbon films was further investigated by employing cross-section transmission electron microscopy, 2-D small-angle X-ray scattering (SAXS), and high-resolution scanning electron microscopy (HR-SEM). Moreover, the transmission electron microscopy analysis of the cross-section of the developed mesoporous carbon films revealed a distorted hexagonal lattice.

Similarly, Ji *et al.* [11] used HRSEM and TEM for characterization of mesoporous graphene framework (MGF) coatings which established that MGF films possessed a 3D ordered, interconnected mesoporous structure with a pore size of ~10 nm (Figure 3.3). The slightly reduced pore size as compared to the initial diameter of Fe_3O_4 nanocrystals (NCs) was attributed to the framework contraction caused by heat treatment. TEM analysis identified the lattice projections such as (111) and (110) which suggested that MGF films possessed the same fcc symmetry inherited from NC superlattice films. For the capacitive performance of MGF films, H_2SO_4 and HNO_3 were used for acid treatment and later FTIR spectra was carried out which indicated a much stronger peak corresponding to OH stretching (broad peak beyond 3000 cm^{-1}) and the significant peak of carbonyl (C=O) extending (~1716 cm^{-1}) after the acid treatment, indicating the presence of a large portion of the hydroxyl (–OH) and carboxyl (–COOH) groups on the graphene framework surface after carrying out the acid treatment.

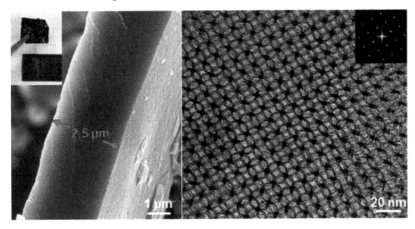

Figure 3.3 Ordered mesoporous few-layer graphene framework films. Reproduced from Reference 11 with permission from American Chemical Society.

3.2.3 Applications

Superhydrophobic free-standing films synthesized by Mittal *et al.* [3] find many applications requiring high temperatures like automobile parts or cooking wares along with capability of long exposure to various corrosive environments. These are even suitable for various applications in microelectromechanical systems, optics along with enhanced resistance to the solvents. Moreover, utilization of MCC for synthesis of superhydrophobic and conductive films assists in release/encapsulation of varying functional materials and tuning the properties like electrical conductivity. Therefore, the carbon nanocontainers embedded in matrices have paved the way for wide range of applications of hydrophobic films based smart coatings.

Fabrication of mesoporous carbon coatings on cordierite, as shown by Wan *et al.* [4], can be used as a reusable adsorbent with high volume processing rates, and recyclability. Similarly, mesoporous carbon films synthesized by spin coating technique by Pang *et al.* [5] find application in sensors, separation processes, membrane based reactors, fuel cells, and many other commercial products. The mesoporous carbon based free-standing thin films, as fabricated by Feng *et al.* [6], have wide range of commercial use in electrochemical and biomedical fields. The films were used to make an electrochemical supercapacitor device which exhibited a capacitance around 136

F/g at 0.5 A/g. Also, nanofilters derived from the above films of carbon demonstrated high rate of filtration of bovine serum albumin and cytochrome c based on size.

Multifunctional superhydrophobic coatings fabricated by Mittal *et al.* [7] based on mesoporous carbon nanocapsules (MCC) with polyvinylidene fluoride (PVDF) find application in bio-medical field with reduced bacterial adhesion. Moreover, the free-standing ordered mesoporous few-layer graphene framework films, synthesized by Ji *et al.* [11], have potential applications in energy storage. The synthesized films were used as electrode materials to build supercapacitors, which exhibited high specific capacitances with excellent cycling stabilities in both aqueous and organic electrolytes, with the capacitive performance comparable to or higher than that of most graphene-based materials.

3.3 Amorphous Carbon Based Coatings

With the improvement in thin film synthesis and development strategies in the last few years such as dip coating, vapor deposition (chemical based), and cathodic arc deposition, fabrication of pure amorphous carbon (a-C) based coatings has been extensively studied. As a result, applications based on amorphous carbon coatings have also enhanced. Figures 3.4 and 3.5 depict the examples of amorphous carbon-coated tin anode material as well as amorphous carbon-coated silicon nanocomposites [14,15]. This section reviews the recent advances in the synthesis, characterization and applications of such functional coatings.

3.3.1 Fabrication

Kim *et al.* [16] coated the $LiNi_{1/3}Mn_{1/3}Co_{1/3}O_2$ cathode material with the amorphous carbon. $LiNi_{1/3}Mn_{1/3}Co_{1/3}O_2$ and sugar were stirred well for 3 hours using ethanol as solvent. The resulting well mixed solution underwent sintering at 350 °C for 1 h in a furnace which resulted in the coating layer being developed. Alakoski *et al.* [17] used the pulsed arc discharge (FPAD) technique to synthesize the diamond-like carbon (DLC) coatings called tetrahedral amorphous carbon (ta-C) coatings on the identical silicon wafers in a vacuum of 100 µPa using high energy unit with high anode-cathode voltage (6 kV), high peak current (13 kA) and short pulses (15 µs). The average carbon ion energies were 600 eV. Due to the large plasma ion energy,

Carbon Based Coatings

low quality of the coatings was observed with sp³ fraction around 40%.

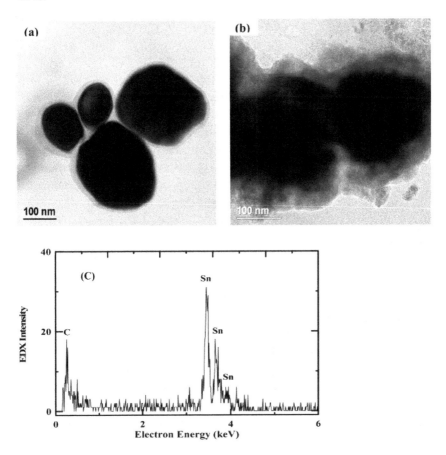

Figure 3.4 TEM images of (a) tin particles and (b) amorphous carbon-coated tin particles, and (c) EDAX spectra of the carbon coated particles. Reproduced from Reference 14 with permission from American Chemical Society.

In another study, Cao *et al.* [18] coated the surface of LiCoO$_2$ with a nanolayer of amorphous carbon using chemical means without reduction of the original surface. Das *et al.* [19] fabricated amorphous carbon based coating by chemical vapor deposition technique on the synthesized Si nanowires using acetylene as the source of carbon. A current density of 12.5 mA cm^{-2} and a working potential of 1.5 kV was used. Also, 0.4 mbar was the working pressure of acetylene and electrode separation was kept at 2.5 cm. A thin layer of a-C was deposited on Si wafers without application of any extra substrate temperature. In addition, Show [20] synthesized a a-C film on Ti bipolar plates at various growth temperatures using radio frequency plasma enhanced chemical vapor deposition (RF-PECVD), where ethylene served as the gas source. Moreover, flow rate of ethylene, RF power and deposition pressure were 10 sccm, 150 W and 0.5 Pa respectively. The growth time was settled for 3 h and the growth temperature was varied between room temperature to 600 °C in order to study the effect of temperature variation.

Liu et al. [21] fabricated carbon coated core shell nanorods of magnetite (Fe_3O_4) using hydrothermal technique with Fe_2O_3 as precursor and citric acid as carbon source. 0.2 g citrus extract was broken down in 5 mL ethanol to produce a homogeneous and transparent mixture. At that point, 0.2 g Fe_2O_3 nanorods was also mixed into the above solution and stirred in a sealed fixed container for 24 h. Drying and grounding were carried out followed by sintering for 2 h at 300 °C. Perez-Huerta and Cusack [22] used amorphous carbon to coat on the surface of two carbonates namely biogenic aragonite and calcite to analyze the effect of the carbon thickness on the crystallographic orientation of the substrate. Carbon coating thickness was controlled utilizing a precision etching-coating system.

3.3.2 Characterization and Testing

Das et al. [19] employed electron microscopy technique to analyze amorphous carbon (a-C) coated silica (Si) nanowires (NWs) and observed that the whole surface of Si NWs was covered with amorphous carbon. Crystalline as well as homogeneous nanowires were observed in the lattice images and same appearance on the coating confirmed its amorphous nature. The FTIR study exhibited an absorbance peak at 2900 cm^{-1} assigned to C–H$_n$ vibrational bonds, which also indicated the presence of amorphous carbon coating on Si NWs. The absorbance peak nearby 1070 cm^{-1} was designated to the Si–O bond obtained from the residual SiO_2 and the one at 3400 cm^{-1} was assigned to O-H bond. Moreover, high vacuum field emission set up was used to study various field emission properties like the enhancement factors and turn-on field which were significantly enhanced due to amorphous carbon coating on the Si NWs.

Show [20] carried out the Raman spectroscopy of the a-C films deposited on the surface of Ti bipolar plates at various growth temperatures and intense Raman signals were observed at around 1410 and

Carbon Based Coatings

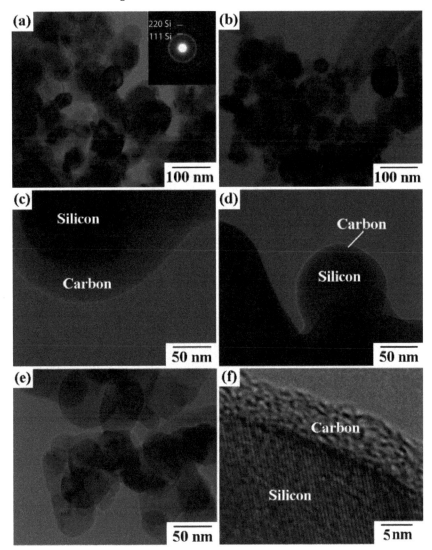

Figure 3.5 TEM images of nanocrystalline Si (a) and carbon-coated Si nanocomposites (b-f). Reproduced from Reference 15 with permission from American Chemical Society.

1560 cm^{-1}, attributed as D and G lines. Sharp D and G lines observed for a-C films deposited at 500 and 600 °C had the peak position displaced to 1350 and 1590 cm^{-1}. Impedance measurement for the fuel cells was also carried out and the Cole-Cole plot exhibited the presence of a curve of semicircular form in the negative impedance (-Z_{img})

region for all cells. Moreover, fuel cells assembled from bipolar plates and coated with a-C film at 600 °C had the bottommost position on Z_{real} axis (real impedance region).

In other studies, Kim et al. [16] used the X-Ray diffraction for the structural analysis of carbon-coated $LiNi_{1/3}Mn_{1/3}Co_{1/3}O_2$. No peaks due to the impurities were seen but the hexagonal crystal structure of α-$NaFeO_2$ with a space group of $R3m$ was visible. DSC was used to measure the thermal stability of the fully charged cell (4.3V versus Li/Li+). Heat release from the uncoated sample was observed to be 161.3 J g^{-1} and the maximum temperature of the exothermic reaction were found as 362.4 and 457.8 °C, respectively. The authors inferred that the layer of coated carbon on the surface of $LiNi_{1/3}Mn_{1/3}Co_{1/3}O_2$ active material suppressed the oxygen release and hence the thermal stability was improved. Kim et al. [16] also evaluated the rate capability of the bare- and carbon-coated powder for which cells underwent charging and discharging at varied cut off voltages. It was observed that the carbon coating improved the rate ability of the material as compared to uncoated material. Moreover, the capacity holding ability of 1 wt% carbon coated material at the rate of 5 °C was found to be 87.4% as compared to that of 0.2 °C rate. Lastly, discharge holding capacity after 50 charge-discharge cycles of same material as above was observed as nearly the same to that of bare material (98% similarity).

Alakoski et al. [17] used the simple visual inspection method to determine the quality of tetrahedral amorphous carbon coating. The authors incorporated the use of X-ray photoelectron spectroscopy and profilometer for determining the sp^3 fraction and coating thickness respectively. It was observed that the tetrahedrally bonded amorphous carbon (ta-C) layer was thickest in the center of the silicon wafer. Cao et al. [18] used various techniques to characterize the carbon coated $LiCoO_2$. From SEM, a very smooth surface of original $LiCoO_2$ was observed. On the other hand, for coated $LiCoO_2$, the surface was found to be rough, non-homogeneous with some small particles found stuck to the surface which seemed like small scattered carbon particles. On the other hand, the TEM analysis exhibited a homogeneously coated nanolayer of carbon on the surface of $LiCoO_2$. X-ray Diffraction patterns were also analyzed for both uncoated and carbon-coated $LiCoO_2$. The main diffraction peaks indicated that both samples had rhombohedral structure. It could also be inferred that there was a weakening and reduction in the intensities of XRD peaks because of amorphous carbon coating on the surface of $LiCoO_2$. The authors also plotted typical Nyquist plots of the coated and the original LiCoO$_2$ composite electrodes. Both the plots exhibited a straight line and a semicircle in low and high frequency regions, respectively.

Liu *et al.* [21] used the high resolution TEM (HRTEM) to characterize the coating of carbon layer on the surface of Fe$_3$O$_4$ nanorods. Moreover, ac impedance spectroscopy, cyclic voltammetry and galvanostatic charge/discharge techniques were employed to evaluate the electrochemical properties of Fe$_3$O$_4$/carbon nanorods as anodes in lithium-ion cells. X-ray diffraction was used to characterize the crystal structure of Fe$_3$O$_4$/carbon nanorods using Cu Kα radiation which indicated that the structure was similar to standard hematite (a-Fe$_2$O$_3$). Moreover, the impurity of α-Fe$_2$O$_3$ phase could not be observed which indicated that the α-Fe$_2$O$_3$ was totally converted to magnetite Fe$_3$O$_4$. On the other hand, TEM analysis of prepared carbon coated Fe$_3$O$_4$ nanorods exhibited the agglomeration of individual nanorods into bundles and all the nanorods had a coating of amorphous carbon. The HRTEM inferred the presence of Fe$_3$O$_4$ lattice and a layer of amorphous carbon having thickness of 2-5 nm. Lastly, energy dispersive X-ray (EDX) study confirmed the presence of amorphous layer of carbon. Perez-Huerta and Cusack [22] utilized electron backscatter diffraction (EBSD) to study crystallographic introduction in biogenic carbonates (aragonite and calcite) in the basic blue mussel, Mytilus edulis, utilizing distinctive sorts of gum and thicknesses of carbon covering. It was observed that there was no appearance of diffraction in aragonite and calcite. EBSD examinations were additionally performed in the electron magnifying instrument in low vacuum mode to analyze the impact of carbon covering thickness with various weight conditions.

3.3.3 Applications

Carbon coated Si nanowires synthesized by Das *et al.* [19] find applications in anti-corrosion coatings. Ti bipolar plates coated with amorphous carbon (a-C) film synthesized by Show [20] are useful for polymer electrolyte membrane fuel cells (PEMFC) as it increased the output power and reduced the internal resistance of the fuel cell. LiNi$_{1/3}$Mn$_{1/3}$Co$_{1/3}$O$_2$ cathode material coated with amorphous carbon is commercially used for lithium batteries because of enhanced electrochemical performance and thermal stability. Hence, it has the potential to be useful for future promising applications in hybrid electric vehicles as an electrode material.

Carbon-coated LiCoO$_2$ lithium ion batteries synthesized by Cao et al. [18] have promising applications where charging and discharging at large rates with high capacity are needed. Carbon coated magnetite (Fe$_3$O$_4$) core-shell nanorods fabricated by Liu et al. [21] are indispensable in ferromagnetic, biomedical, and catalysis applications. Perez-Huerta and Cusack [22] generated the coating on biogenic carbonates which are used as impregnating agents in the resins like epoxy resin for grinding and polishing applications.

3.4 Coatings Based on Pyrolytic Carbon and Carbon Black

Some of the characteristics of pyrolytic carbon resemble with that of graphite, but different in the sense that it has covalent bonding present on its surface in the graphene sheets because of various defects and imperfections generated during its manufacturing. Pyrolytic carbon has vast range of applications in automobiles, nuclear reactors, and spectrometry. Its extensive use in biomedical applications has made it very relevant in current perspective. On the other hand, carbon black is obtained from the combustion of various types of petroleum products. Properties like high surface to volume ratio make carbon black very appealing for research and development. The below section summarizes some of the carbon coatings synthesized using the two materials of pyrolytic carbon and carbon black.

3.4.1 Fabrication

Kim et al. [23] synthesized carbon coating layers comprising of silicon carbide and pyrolytic carbon on UO$_2$ pellets by utilizing combustion reaction technique between silicon and carbon. For the coating purpose, propane underwent thermal decomposition reaction in a chemical vapor deposition unit at 1250 °C and microwave pulsed electron cyclotron resonance plasma enhanced chemical vapor deposition (ECR PECVD) at 500 °C by utilizing silane. The carbon deposition unit comprised of a 1500 kW resistance heater, a preheater of the source gasses, a gas blender, a temperature controller, and a gas stream control framework.

Lopez-Honorato et al. [24] fabricated silicon carbide and pyrolytic carbon coatings using time domain thermoflectance on fuel particles (simulated). Three variety of samples were synthesized utilizing fluidized bed chemical vapor deposition. Two single layer PyC coatings and a triple layered coating comprising of a buffer, silicon carbide

(SiC) and inward pyrolytic carbon (IPyC) were spread on alumina particles 250 μm in radius. The single layer coatings were deposited with 50% v/v acetylene concentration at 1250 °C (1250-PyC) and 1450 °C (1450-PyC). For the triple layered particle, a low density PyC (buffer) was synthesized with 40% v/v acetylene at 1450 °C, while the inward high density PyC was deposited with a blend of 33% v/v propylene/acetylene at 1300 °C.

In another study, Kim et al. [25] fabricated coatings using carbon black on the surface of $LiCoO_2$ cathode and utilized surfactants for high-density Li-ion cell. The coating strategy contained two stages: (i) scattering of amassed carbon black utilizing orotan®, a polyacrylate dispersant; (ii) carbon-black covering of the cathode material utilizing a surfactant gelatin which is amphoteric in nature. The procedure diminished the carbon content in the electrode and had no effect on the cycle-life working and efficiency of the cell. Furthermore, it also enhanced the rate ability of the Li-ion particle cell. For coating purpose, around 2 g of gelatin was mixed in 30 ml of DI water for 10 min at 50 °C. Finally, the carbon-dispersed orotan® solution was added in the above mixture and gelatin coating was accomplished on the dispersed carbon black. The pH of gelatin was required to be brought close to its isoelectric point of pH ~ 4–5 by utilizing acetic acid, as it was observe to help in the direct coating of carbon on the cathode surface.

In a related study, Ponrouch et al. [26] fabricated carbon coatings by physical deposition of carbon achieved by dissipation under vacuum (around 10^{-4} mbar) performed at normal room temperature along with dry conditions. A commercially available carbon evaporator was used for the coating purpose on the electrode active material powders. A specimen holder was fabricated for accomplishing the uniform coverage on all particles and comprised of a glass petri dish fixed to the traditional pivoting/tilting plate of the device. For achieving a uniform and smooth coating, the persistent mixing of the sample along with rotation by using the sample holder was carried out in the study.

In another study, Rodriguez-Mirasol et al. [27] also formed pyrolytic carbon within a microporous zeolite template using chemical vapor infiltration, as shown in Figure 3.6. Propylene, in a nitrogen stream, was used as the carbon precursor. The process resulted in the microporous carbon structure with high surface area, along with wide microporosity, well-developed mesoporosity and high adsorption capacity.

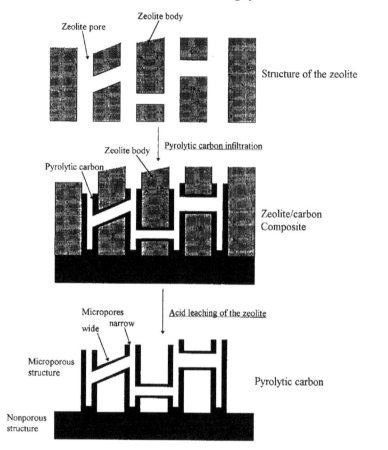

Figure 3.6 Scheme of pyrolytic carbon infiltration of zeolite, followed by removal of the zeolitic substrate. Reproduced from Reference 27 with permission from American Chemical Society.

3.4.2 Characterization and Testing

Kim et al. [23] carried out the chemical analyses of the multi-layer coating with X-ray diffraction and Auger electron spectroscopy (AES) along with TEM and SEM. SEM exhibited uniform layers of coating having silicon carbide and pyrolytic carbon due to the combustion reaction. The X-ray spectra indicated β-SiC was the product with negligible presence of unreacted silicon. Moreover, AES and XRD studies confirmed the presence of pyrolytic carbon and silicon carbide at the inner and outer coating layers respectively.

Carbon Based Coatings

Lopez-Honorato et al. [24] used the time-domain thermoreflectance for pyrolytic carbon and silicon carbide coatings on fuel particles and performed thermal conductivity mapping to obtain the line profiles of the coatings. Later, SEM of these coatings was carried out which led to the inference of differences between density and porosity. Values of thermal conductivity obtained from the maps for PyC were 4.2 W/m K for 1250-PyC and 3.4 W/m K for 1450-PyC and corresponding density values were 2.12 and 1.41 g/cm^3 respectively.

Kim et al. [25] carried out the studies of volumetric limits in Li-ion particle cells utilizing uncovered or carbon-coated LiCoO$_2$ electrodes. Discharge capacities of LiCoO$_2$ were seen from the anode/cathode dimensional proportion. To further study the impact of surfactant on the stability of the Li cell, the cells having bare or carbon-covered cathodes were charged to 4.2 V after each 50 cycles and the changes in thickness of the cell at room temperature was calculated. It was concluded that the cells displayed comparable swelling with expanding the number of cycles.

Ponrouch et al. [26] carried out TEM examination to analyze the carbon deposition on the electrode materials. TEM demonstrated a clear carbon layer at the surface of the dynamic material particles with a thickness ranging from 1 nm up to 4 nm. Additionally, TGA for carbon covered NTO powder with a layer of carbon having thickness around 2.5 nm were analyzed. Weight loss due to the decay of the carbon layer into CO and CO_2 was observed. Raman spectroscopy was utilized to get further knowledge into the graphitization level of the deposited carbon layer. Apart from the sharp crest at around 1000 cm^{-1} and the more extensive and less visible peak near 1100 cm^{-1}, the range comprised of a several broad bands which overlapped on one another, the majority of which were related to C vibrations. The effect of carbon coating on the kinetics of electrolyte reduction and SEI development was examined through the investigation of Co_3O_4 based electrodes. Electrodes synthesized with uncoated and carbon coated nanosized Co_3O_4 were cycled in galvanostatic mode with potential restriction (GCPL) with a progression of 10 cycles.

3.4.3 Applications

The multilayer coatings of pyrolytic carbon and silicon carbide over the surface of UO_2 pellets synthesized by Kim et al. [23] represented the coating process which is beneficial to study the feasibility of the above coatings for fuels of light water and heavy water reactors

(LWR, HWR). Lopez-Honorato et al. [24] also developed pyrolytic carbon and silicon carbide coatings on simulated fuel particles were helpful for nuclear purpose. Plotting of the conductivity of coated fuel particles provided valuable information to optimize fuel efficiency amid the operation of atomic reactors. Direct carbon-black coating of the cathode material using a surfactant fabricated by Kim et al. [25] improved the rate capacity of the Li-ion cell and, thus, has excellent application in technology like mobiles, PCs, cameras, laptops, etc.

Carbon coatings synthesized by Ponrouch et al. [26] are largely utilized to fabricate conductive carbon coatings with a thickness of ca. 1 nm and upwards on insulating specimens to empower charge free imaging in electron microscopy and microprobe examination. Also, it has the benefit of delivering coatings covering uneven and non-homogeneous surfaces.

3.5 Coatings Based on 'Carbon Alcohols' as Source

Various coatings have been synthesized in the recent past with carbon based alcohols as source like ethylene glycols, ethyl alcohol, ethanol, etc. This section summarizes the fabrication techniques and performance results of such coatings.

3.5.1 Fabrication

Lin et al. [28] synthesized carbon coating using ethylene glycol as source using the substrate $LiFePO_4$, with various molar ratios of high electron conductive iron phosphide phase by an aqueous sol–gel technique in a reductive sintering environment. Different parameters were utilized for modifying the microstructure and varying phase concentration of products. The $LiFePO_4$ cathode material was prepared using the sol-gel method utilizing various constituents as $FeC_2O_4 \cdot 2H_2O$ (ferrous oxalate), $LiNO_3$ (lithium nitrate) and $NH_4H_2PO_4$ (dihydrogen ammonium phosphate). In another study, Klebanoff et al. [29] utilized ethyl alcohol as the carbon source for manufacturing radiation-instigated protective carbon coating on extraordinary bright optics. A 5 Angstrom carbon coating was kept on EUV Mo/Si optics by means of co-exposure to radiation (EUV photons, electrons) and ethanol vapor. A 2 kV electron beam was used as a EUV/water exposure. Moreover, a sputtered carbon sample was used and prepared in a custom-built magnetron sputtering system which was operated at the base pressure of less than $3*10^{-7}$ Torr using argon at a

Carbon Based Coatings

constant pressure of 0.9 mTorr as a sputtering gas. As an experimental setup, two substrate holders were spun during the entire deposition process.

Wang et al. [30] fabricated a nano-carbon coating on the surface of LiFePO$_4$ using ethanol as carbon source using spray pyrolysis system. In this study, effect of corrosion was studied by the addition of carbon layer at olivine LiFePO$_4$ stored in moisture-contaminated electrolyte. Effect of iron dissolution on corrosion was also studied.

3.5.2 Characterization and Testing

Lin et al. [28] studied the release/discharge abilities to analyze the impact of the carbon covering and iron phosphides on the electrochemical properties of the LiFePO$_4$/C electrodes at rates of 0.1-5C (1C = 170mAhg^{-1}) and examining the CV bends. It was observed that carbon covering in a ratio of 1.5 wt% significantly diminished the molecule size of LiFePO$_4$, and enhanced the rate ability of LiFePO$_4$. Additionally, the impact of the weight of FeP on the capacity of the carbon coated LiFePO$_4$ varied at various release rates. Incrementing the concentration of FeP from 1.2 to 3.7 wt% somewhat diminished the capacity of LiFePO$_4$/C at low release rate (0.1C and 1C). Likewise, an excessive amount of iron phosphides brought down the release limit of the electrode since these were not reactive for the disinsertion/insertion of lithium particle. XRD patterns of the products synthesized by various blend parameters demonstrated that the principle phase was LiFePO$_4$, along with iron phosphides peaks. Additionally, SEM analysis of the LiFePO$_4$ affirmed that the molecule size of the carbon free LiFePO$_4$ synthesized from gel without ethylene glycol was substantially bigger than that of the LiFePO$_4$ particles coated with carbon.

Klebanoff et al. [29] carried out the sputter Auger depth profiling, Auger electron spectroscopy, and EUV reflectivity calculations of the synthesized radiation-induced carbon coating on EUV optics. It was observed in a nutshell that the coating was void free and protected the optic from water-induced oxidation at the water partial pressures which were used in the experiment. Moreover, the coating was resistant to atmospheric degradation along with gasification with a 0.5% reduction in the relative reflectivity of the optic.

Wang et al. [30] used various testing methods to describe the morphology and structure of olivine LiFePO$_4$ after the application of nanocarbon coating and iron dissolution. The SEM images of carbon coated LiFePO$_4$ showed that the conductivity was greatly improved after coating and a homogeneous layer of nanocarbon was visible. Using FIB along with HRTEM, the thickness was determined to be 10 nm and a clear Au protection film was also observed. Also, the corrosion extent of coated LiFePO$_4$ was greatly reduced after aging process, along with reduced loss of iron, hence, surface corrosion was greatly reduced. Further, ToF-SIMS testing was performed for coated LiFePO$_4$ which showed much less corrosion degradation by acidic species during aging process, as carbon coating was helpful in preventing attack from HF on LiFePO$_4$ surface and hence provided a protective layer. Lastly, X-ray absorption near-edge structure (XANES) spectra was studied for coated LiFePO$_4$ and was found to be nearly same for uncoated sample, except indication of partial oxidation of iron. On the whole, materials stability and protection was enhanced by addition of carbon coating based on ethanol on the surface of LiFePO$_4$.

3.5.3 Applications

The carbon coating on LiFePO$_4$ prepared by Lin *et al.* [28] find enormous applicability in increasing the intrinsic electronic conductivity, hence, making it a promising future commercial alternative cathode material for lithium ion battery. Radiation-induced protective coating fabricated by Klebanoff *et al.* [29] are commercially viable as these are resistant to atmospheric degradation. Moreover, these coatings improve the properties of EUV optics which are extremely important for the manufacturing of next-generation semiconductor chips.

The nanocarbon coatings synthesized by Wang *et al.* [30] on the surface of olivine LiFePO$_4$ greatly enhanced its performance and, hence, make it a potential candidate as a safe and efficient cathode material for advanced LIBs applied in commercial electric vehicles, representing a next generation safe technology for future.

3.6 Coatings Based on Graphene and Other Related Materials

Graphene has been extensively used in the recent years for various research studies. As a result, graphene based materials have been used in a wide variety of protective coatings. For instance, Figure 3.7 shows the partially covered platinum surface with graphene [31]. The graphene coated surface exhibited effective protection against CO and O$_2$. Another example of graphene based surface protection is

Carbon Based Coatings   93

also demonstrated in Figure 3.8 for Gr/SiO$_2$, Cu, and Gr/Cu samples, where graphene was coated on the surface for corrosion protection [32].

The below section summarizes the recent studies in which graphene and other related functional nanomaterials have been used to synthesize the coatings with tuned structure and properties, along with their performance in diverse areas.

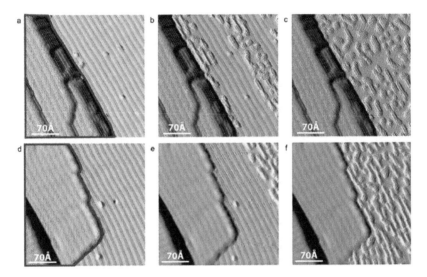

Figure 3.7 STM image of a Pt surface partially coated by graphene (the area at the right side is uncoated, and the rest is coated by graphene). (a-c) Surface exposed to 0, 3, and 63 L of CO, respectively. (d-f) Coated surface exposed to 0, 25, and 40 L of O$_2$, respectively. Reproduced from Reference 31 with permission from American Chemical Society.

3.6.1 Fabrication

Nishihara *et al.* [33] covered the pore surface of mesoporous silica SBA-15 with 2,3-dihydroxynaphthalene (DN) through a lack of hydration response behavior between the surface silanol aggregates in SBA-15 and the hydroxyl groups of the DN particles. Afterwards, the carbonization of DN in the SBA-15 pores brought about to a great degree thin carbon layer arrangement in the pores including 1-2 graphene sheets. For the synthesis process, the authors added 2.3 g of DN in 5 ml of acetone and further dissolved the mixture into 0.47 g of dried SBA-15 in vacuum. The blend was mixed for a few hours at room temperature which led to the evaporation of acetone. The subsequent strong blend of DN and SBA-15 was warmth treated at 573 K for 1 h under N_2 stream. Amid this progression, fluid state DN was permitted to undergo reaction with SBA-15 through a dehydration reaction between the silanol groups on the pore surface of SBA-15 and the hydroxyl aggregates of DN. The unreacted DN was then washed away with acetone to acquire DN-covered SBA-15 (DN/SBA-15). Lastly, the synthesized DN/SBA-15 was warmth treated under N_2 stream at 1073 K for 4 h to carbonize DN in the silica pores. Thus, graphene was covered/coated on silica particles.

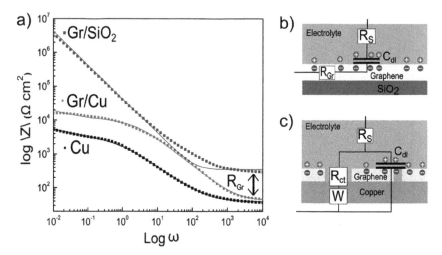

Figure 3.8 (a) Bode plots of Gr/SiO_2, Cu, and Gr/Cu samples (shown as solid symbols), whereas the lines represent best fits to the equivalent circuit models; (b) equivalent circuit model used in modeling Gr/SiO_2 devices, and (c) equivalent circuit model for Cu and Gr/Cu devices. Reproduced from Reference 32 with permission from American Chemical Society.

In another study by Yoshida *et al.* [34], carbon film was generated on SiC filaments in the fabric by electrophoretic deposition technique utilizing industrial colloidal graphite homogeneous mixture and was compared with carbon coating on SiC strands in fiber material synthesized by dip coating or vacuum penetration utilizing the above prepared suspension of colloidal graphite. Lee and Lee [35] synthesized the carbon coating on nano-silica particles. The carbon coating comprised of graphite and coal–tar pitch. For synthesizing oxides of nano-Si, mechano-chemical reduction of SiO was carried out by Al.

Later, carbon coating was performed on the ball milled mixture of above oxides with graphite using pyrolysis/heating of coal tar at approx. 900 °C. The procedure was carried out in a glove container with argon gas. Later, size and shape of the composite was controlled by grounding the coated carbon product using rotor mill using variable speeds.

In a recent study by Li and Zhang [36], superhydrophobic coatings were fabricated by using spray-coating technique. The uniform suspension of polysiloxane/multiwalled carbon nanotubes (POS/MWCNTs) nanocomposites was spray coated onto various substrates which was followed by curing for a fixed time span. N-hexadecyltriethoxysilane and tetraethoxysilane underwent hydrolytic co-condensation with MWCNTs which was activated by acid. In another study, Sevastyanov et al. [37] fabricated conducting coatings based on carbon nanotubes and other carbon nanostructures and SnO_2. Vaporized transport method was employed for obtaining the coatings. The accompanying nanomaterials utilized in the study were: carbon nanotubes (CNT), carbon nanofibers (CNF) and carbon nanoflakes (CNFL). For coatings purpose, nanomaterials were used to prepare corresponding suspensions. Carbon nanostructure layers were laid on 10*10 mm glass substrates; cleaned silicon plates were utilized as an extra reference samples. The suspensions with carbon nanomaterials were placed in an ultrasonic tank for 30 min and set into the assembly of a vaporized generator quickly before deposition.

Wang et al. [38] also reported the graphene-sulfur composites by wrapping polyethylene glycol (PEG) coated sulfur particles with mildly oxidized graphene oxide sheets, which were decorated by carbon black nanoparticles (Figure 3.9).

3.6.2 Characterization and Testing

Nishihara et al. [33] utilized different testing strategies to describe the carbon covered mesoporous silica. TGA was performed to decide the thermal response of DN and carbon stacked on SBA-15. The silanol aggregates on the silica surface were additionally examined with a Fourier transform infrared (FT-IR) spectrometer in the absorbance mode. Powder X-beam diffraction (XRD) was also carried out and the surface areas were ascertained utilizing the Brunauer-Emmet-Teller (BET) technique. The FT-IR spectra of SBA-15 and DN/SBA-15 demonstrated two wide peaks at around 810 and 1030 cm^{-1} which suggested crucial silicon-oxygen vibrations. After reaction with DN, a few new peaks arising from DN were observed, for example, C=C extending vibrations at around 1500 cm^{-1} and C–H out-of-plane disfigurement vibrations at around 750 cm^{-1}. The XRD patterns of SBA-15 and carbon/SBA-15 were observed to be of same nature which demonstrated that the long-run consistency of the silica system in SBA-15 was retained prior and after the carbonization procedure. Additionally, SEM of the initial and carbon-stacked SBA-15 (carbon/SBA-15) demonstrated no morphological change even after the carbon covering, which led to the inference that the carbon was consistently present on the internal surface of the SBA-15 mesopores. In conclusion, the BET analysis concluded that the mesopores were infinitesimally present even on carbon/SBA-15.

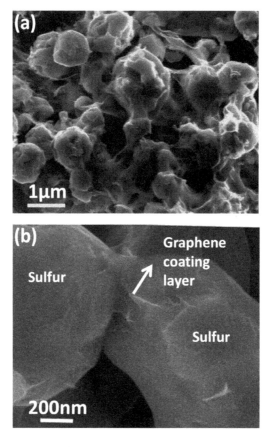

Figure 3.9 SEM images of graphene-sulfur composite at low (a) and high (b) magnifications. Reproduced from Reference 38 with permission from American Chemical Society.

Yoshida et al. [34] carried out the SEM analysis of carbon-coated SiC fibers generated by various techniques like dip-coating, electrophoretic deposition, and vacuum infiltration. It was observed that the fiber was polycrystalline, and consisted of fine β-SiC particles. In the case of coating with electrophoretic deposition, flaky particles of graphite totally coated the surface of SiC fibers. However, in the case of the coatings by dip coating and vacuum filtration methods, only partial presence of graphite particles on SiC fibers was observed. Later, the analysis of formed composites was performed, especially the bending test and load-crosshead displacement plots of composites were analyzed at room temperature. Very similar load-crosshead curves were obtained for coating with electrophoretic deposition and vacuum filtration methods. Also, it was observed that as crosshead displacement increased after reaching a maximum valued of load, a decrease in the load, though gradually, was observed. However, in the case of coating using dip coating method, the load indicated a linear relationship with crosshead displacement and afterwards a sharp and sudden decrease. The authors concluded that the carbon coating was homogenous.

In the study by Lee and Lee [35], XRD technique was used for investigating the carbon-coated nano-Si dispersed oxides/graphite composite. The SiO-Al-Li_2O_2 powder mixture was milled for different durations, followed by XRD analysis. It was inferred that initially without milling, the pattern had a weak broad signal. Moreover, Si peaks appeared after 6 hours of milling with the disappearance of Li_2O_2 diffraction peaks. It was also seen that as the milling time was increased to 15 hours, there was an increment in the intensity of Si diffraction peaks. The peak of Al_2O_3 was also observed with negligible effect of residual SiO and Al on its electrochemical properties. Calculation of discharge capacity as a function of cycle number for varying milling times was performed for the composite electrode. On the whole, authors inferred that the carbon coating treatment on the composites greatly enhanced their cyclic ability along with minimized irreversible capacity.

Li and Zhang [36] used a wide variety of techniques to characterize the nanocomposites and the coatings. Sand abrasion and water jetting techniques were used for healing capability tests of POS/MWCNTs. In addition, FTIR, SEM, TEM, X-ray photoelectron spectroscopy (XPS) of samples was performed. TEM of nanocomposite exhibited no silica nanoparticles. Moreover, the POS/MWCNTs surface which was damaged because of the artificial sand abrasion and water jetting were healed with the help of toluene and curing. From the FTIR and XRD, it was inferred that the coatings were superhydrophobic, had large contact angles, ultralow sliding angles and were very stable. Furthermore, there was a minute change in CAs of coatings and low SAs were maintained after exposing to severe stability tests of sand abrasion.

Surface of the conducting coatings synthesized by Sevastyanov *et al.* [37] based on different carbon nanomaterials was investigated using various techniques to study composition and structure like X-ray phase analysis (XPA), TEM, SEM, AFM, etc. From SEM, a coating was observed to be deposited on the surface of CNFL sample, along with the presence of voids. However, for CNF-SnO_2, a different coating having polycrystalline tin oxide was observed on its surface which covered the earlier formed layer of carbon nanostructures on the coating surface. Moreover, XRD analysis indicated that a tin dioxide coating was present on all samples.

3.6.3 Applications

The carbon coatings synthesized by Nishihara *et al.* [33] on mesoporous silica find their applications commercially as materials like molecular sieves, adsorbents, etc. Due to the increased hydrophobicity and electrical conductivity, coated silica finds commercial use in solutions which are ionic and organic, and is generally used as supercapacitors electrodes. On the other hand, carbon coatings fabricated by Lee and Lee [35] on nano-silica proved to be a potential anode material for lithium ion batteries and for electronic and ionic conduction.

Hydrophobic coatings fabricated by Li and Zhang [36] are ultrastable, healable and are commercially feasible in various applications like separation of oil/water, self-cleaning materials, and controlled drug release. Furthermore, these POS/MWCNTs coatings can be applied onto various substrates like cellulose paper, PTFE, polymeric and aluminum plates along with polymer based textiles, etc. Also, the conductive coatings synthesized by Sevastyanov *et al.* [37] were useful for photovoltaic cells especially organic semiconductors and for the production of solar cells.

3.7 Miscellaneous Carbon Coatings

In the recent past, other carbon based compounds have also been used to formulate functional coating structures, like carbon xerogel,

Carbon Based Coatings

microporous carbon, carbides, etc. The below section summarizes the research studies using such materials for the generation of functional coatings.

3.7.1 Fabrication

Cushing and Goodenough *et al.* [39] carried out the carbon coating on the surface of α-NaFeO$_2$-structured LiMn$_{0.5}$Ni$_{0.5}$O$_2$ cathode materials. Carbon xerogel was used as the source of carbon, which was derived from resorcinol-formaldehyde (R-F) polymer. LiMn$_{0.5}$Ni$_{0.5}$O$_2$ particles were carbon-coated by combining as-prepared LiMn$_{0.5}$Ni$_{0.5}$O$_2$ and R-F in an 85:15 weight ratio, compacting the mixture into a 2.5 cm diameter pellet, and heating the pellet to 600 °C for 6 h in air. In a study by Chmiola *et al.* [40], etching of supercapacitor electrodes into conductive titanium carbide substrates was carried out which resulted into the monolithic carbon films. For this purpose, cutting of sintered TiC ceramic plates was carried out and the plates were later polished to reduce the thickness to around 300 nm so that resistance can be minimized before chlorination process starts on each face ranging in a time span of 15 sec to 5 mins. Finally, the obtained coatings had the thickness ranging from 2 to 200 mm. Hence, the technique successfully produced thin carbide derived carbon (CDC) films on bulk TiC ceramic plates.

A recent study by Chung and Manthiram [41] showcased the microporous carbon (MPC) coating on the polyethylene glycol (PEG) as a polysulfide trap. The MPC/PEG slurry was synthesized by mixing 80 wt% MPCs with 20% PEG in 3 mL of isopropyl alcohol (IPA). Later, slurry coating was performed on one side of Celgard separator using tape casting method via an automatic film applicator with blade (standard number 1), followed by drying. Roldan *et al.* [42] formulated a hydrothermal carbon (HC) layer coating on graphite microfiber felts using one-pot synthesis method. For this purpose, piece of graphite felt was immersed into the carbohydrate solution in an autoclave vessel. Finally, after washing and drying with ethanol and water, the carbon felt got coated with hydrothermal carbon layer denoted by HC-CF. Later, using concentrated sulfuric acid, the treated felt was subjected to sulfonation and the treated sample was denoted as HO$_3$S-HC-CF.

Zhu *et al.* [43] coated the Li$_4$Ti$_5$O$_{12}$ (LTO/C) nanocomposite particles with carbon and proposed a model to illustrate the necessity of the carbon content optimization in the composite material. For the coating purpose, varying amount of glucose in $TiOSO_4$ aqueous solution was loaded in an autoclave at 180 °C for 6 h. The resulting precipitate was filtered and washed, followed by drying to generate TiO_2/C precursor. It was later mixed with $LiOH.H_2O$, followed by calcination. In another recent study by Zhang et al. [44], carbon coated Fe_3O_4 nanospindles were synthesized by partial reduction of monodispersed hematite nanospindles. For the fabrication purpose, forced hydrolysis of ferric chloride solution along with addition of phosphate ions was carried out which led to the synthesis of hematite particles followed by the coating of carbon layers by pyrolysis of glucose under hydrothermal conditions. Later, drying and heating was performed under N_2 environment so as to achieve carbonization of the layers. Lastly, partial reduction of inner hematite spindles was carried out to generate magnetite, which resulted in final Fe_3O_4-C nanocomposites.

In another study, Wang et al. [45] derived interconnected carbon nanosheets from hemp and used for ultrafast supercapacitors with high energy. Figure 3.10 also demonstrates the morphological features of the interconnected partially graphitic carbon nanosheets.

3.7.2 Characterization and Testing

Cushing and Goodenough [39] characterized the products using XRD with a Cu-Kα radiation. Elemental analysis was studied using absorption spectrophotometer by utilizing the method of standard addition. Moreover, the carbon content of $LiMn_{0.5}Ni_{0.5}O2$-C (LMNO-C) was determined by TGA carried out in 30-800 °C range in oxygen atmosphere. XRD exhibited $LiMn_{0.5}Ni_{0.5}O_2$ (LMNO) as O_3 structure of $LiNiO_2$ and the presence of nickel and manganese ions was confirmed. Also, the elemental analysis of LMNO-C indicated the presence of 2.5% (w/w) of carbon. Moreover, SEM images of LMNO and LMNO-C confirmed the presence of small particles in 10-100 nm size range which were bound in the form of agglomerates. On the other hand, EDX carbon map of LMNO-C showed that carbon was uniformly distributed throughout the agglomerates. Also, the discharge capacities were measured by cycling the coated cathodes between 2.75 V and 4.25 V. The values of 87, 101 and 108 mAh/g were obtained corresponding to current densities of 1, 0.5 and 0.25 mA/cm^2.

Chmiola et al. [40] carried out the SEM analysis of the carbide based carbon films on TiC which exhibited an excellent adherence on TiC surface along with uniform thickness and some microcracking

Carbon Based Coatings

due to the presence of tensile stress because of Ti removal. Raman spectroscopy further verified the findings of SEM analysis and it was inferred that an increment in the microstructural ordering at the surface of the thicker films was present. Furthermore, volumetric capacitance of $TEABF_4$ and H_2SO_4 was also calculated for different thickness of the films and it was observed to decrease with increased thickness of the coatings. Chung and Manthiram [41] carried out the SEM analysis of microporous carbon (MPC) coating on the polyethylene glycol (PEG). It was observed that MPCs formed clusters of porous nature with size in microns range and were attached to the separator. Moreover, the MPC nanoparticles had large surface area and

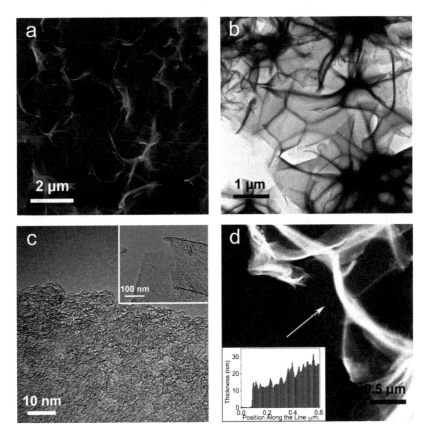

Figure 3.10 (a&b) SEM and TEM micrographs of the interconnected 2D structure; (c) high resolution TEM micrograph indicating the porous and partially ordered structure and (d) ADF TEM micrograph and EELS thickness profile (inset). Reproduced from Reference 45 with permission from American Chemical Society.

high pore volume, along with many micropores. It was also concluded that polysulfides were intercepted by the microporous trapping sites of the MPC/PEG coating. BET analysis, on the other hand, indicated that the coatings were very effective as electrolytes were absorbed in the porous space. Lastly, electrochemical analysis of the Li-S cells was performed with the fabricated MPC/PEG coated separator

Roldan et al. [42] characterized the coated layer of hydrothermal carbon on the graphite felt using SEM, IR-spectroscopy, XPS, etc. and studied its morphology and surface chemistry. Esterification technique was also utilized to test the coatings after functionalization with sulfonic acid groups. From SEM, it was observed that the thickness of the layer increased as the concentration of the glucose enhanced. It was also inferred that hydrothermal carbon synthesized by using HNO_3 treated carbon felt yielded higher carbon sphere than the one synthesized by carbon felt without HNO_3 treatment. The functional group analysis exhibited various bands after the coating with hydrothermal carbon with most prominent one at 1710 cm^{-1} of C=O and at 2900 and 3000-3700 cm^{-1} representing methyl/methylene and hydroxyl/carboxyl groups respectively [46-49].

Zhu et al. [43] studied the carbon coating by in-situ Raman analysis. Moreover, structural and morphological studies of the samples were performed by XRD and SEM. Content of carbon in the coatings was studied by elemental analyzer and TGA. Electrochemical impedance spectroscopy (EIS) was also studied at 10 mV AC Volts between 100 kHz to 100 mHz. XRD analysis of TiO_2 precursor and the LTO/C samples exhibited no carbon peaks because of the amorphous nature of carbon and its low content. The TEM analysis confirmed the results of XRD and it was further concluded that lower temperature of calcination favors formation of smaller nanoparticles with diameter varying from 70 to 220 nm. Moreover, Raman spectra showed a strong D band, thus, again validating the amorphous nature of carbon with multiple defects. A G band was also noticed corresponding to graphitized carbon.

Zhang et al. [44] investigated the properties of carbon coated Fe_3O_4 nanospindles using SEM, TEM, XRD and various electrochemical techniques. SEM images of nanospindles alone exhibited tiny crystals with a diameter of about 106 nm. Also, SEM images of carbon coated nanospindles showed uniform coating layers. On the other hand, SEM images of final Fe_3O_4-C nanocomposites were smoother with small void spaces. Furthermore, the TEM images confirmed the presence of the outer layer of carbon and inner magnetite core. To determine the pore structure of Fe_3O_4-C, BET analysis was carried out. The surface area was approximated to be 35.1 m² and the Barrett-Joyner-Halenda (BJH) pore size distribution indicated presence of mesopores in the composite, which led to the better cyclic performance for lithium batteries.

3.7.3 Applications

The carbon coating generated by Cushing and Goodenough [39] on the surface of $LiMn_{0.5}Ni_{0.5}O_2$ cathode using xerogel as carbon source is commercially useful in rechargeable lithium ion batteries. On the other hand, the carbon coatings fabricated by Chmiola *et al.* [40] using carbide are useful for micro-supercapacitors as these provide a framework for their integration into a variety of devices because of improved functionality and reduced complexity. The carbon coating led to reducing the abrupt failure and increasing the cycle life of the micro-batteries.

The microporous coating carried out by Chung and Manthiram [41] on the surface of separator having polysulfide trap is beneficial for lithium-sSulfur batteries due to high energy density and discharge capacity. The hydrothermal carbon coating carried out by Roldan *et al.* [42] on the graphite felts was applicable for acid catalyst applications. Also, the graphite felt's electrical conductivity was increased which makes its use more widespread for electrochemical devices. The macroporous structure of felt also made it commercially applicable in flow chemistry studies.

The carbon coating formulated by Zhu *et al.* [43] on the surface of $Li_4Ti_5O_{12}$ (LTO/C) nanocomposite particles paved way for high energy density and power batteries based on lithium ions which made them more efficient for electrochemical energy storage. In a very similar fashion, coatings fabricated by Zhang *et al.* [44] on Fe_3O_4 nanospindles served as a valuable anode material for lithium ion batteries because of enhanced performance and further paved the way for upcoming lithium ion batteries with large energy capacities.

3.8 Conclusions

Carbon based coatings/films have been synthesized in the recent years for various applications like reducing the effect of corrosion on the metallic surfaces, improving properties of lithium ion batteries, etc. Carbon has been used widely because of the multiple properties it exhibits like hybridization, existence in different forms like graphene, nanotubes, amorphous carbon, etc. Major advances have been achieved recently in the synthesis of carbon based films and coatings where different forms of carbon functionalization have been performed, which have further enhanced the commercial potential of these materials.

References

1. Broitman, E., Neidhardt, J., and Hultman, L. (2008) Fullerene-like carbon nitride: A new carbon-based tribological coating. In: *Tribology of Diamond-Like Carbon Films: Fundamentals and Applications*, Donnet, C., and Erdemir A., eds., Springer, USA, pp. 620-653.
2. Broitman, E., Czigany, Z., Greczynski, G., Bohlmark, J., Cremer, R., and Hultman, L. (2010) Industrial-scale deposition of highly adherent CNx films on steel substrates. *Surface and Coatings Technology*, **204**, 3349-3357.
3. Mittal, N., Deva, D., Kumar, R., and Sharma, A. (2015) Exceptionally robust and conductive superhydrophobic free-standing films of mesoporous carbon nanocapsule/polymer composite for multifunctional applications. *Carbon*, **93**, 492-501.
4. Wan, Y., Cui, X., and Wen, Z. (2011) Ordered mesoporous carbon coating on cordierite: Synthesis and application as an efficient adsorbent. *Journal of Hazardous Materials*, **198**, 216-223.
5. Pang, J., Li, X., Wang, D., Wu, Z., John, V. T., Yang, Z., and Lu, Y. (2004) Silica-templated continuous mesoporous carbon films by a spin-coating technique. *Advanced Materials*, **16**, 884-886.
6. Feng, D., Lv, Y., Wu, Z., Dou, Y., Han, L., Sun, Z., Xia, Y., Zheng, G., and Zhao, D. (2011) Free-standing mesoporous carbon thin films with highly ordered pore architectures for nanodevices. *Journal of the American Chemical Society*, **133**, 15148-15156.
7. Mittal, N., Kumar, R., Mishra, G., Deva, D., and Sharma, A. (2016) Mesoporous carbon nanocapsules based coatings with multifunctionalities. *Advanced Materials Interfaces*, **3**(10), 1500708.
8. Yang, Y., Yu, G., Cha, J. J., Wu, H., Vosgueritchian, M., Yao, Y., Bao, Z., and Cui, Y. (2011) Improving the performance of lithium-sulfur batteries by conductive polymer coating. *ACS Nano*, **5**, 9187-9193.
9. Yang, S. J., Nam, S., Kim, T., Im, J. H., Jung, H., Kang, J. H., Wi, S., Park, B., and Park, C. R. (2013) Preparation and exceptional lithium anodic performance of porous carbon-coated ZnO quantum dots derived from a metal-organic framework. *Journal of the American Chemical Society*, **135**(20), 7394-7397.
10. Zhao, X. Y., Tu, J. P., Lu, Y., Cai, J. B., Zhang, Y. J., Wang, X. L., and Gu,

C. D. (2013) Graphene-coated mesoporous carbon/sulfur cathode with enhanced cycling stability. *Electrochimica Acta*, **113**, 1 256-262.

11. Ji, L., Guo, G., Sheng, H., Qin, S., Wang, B., Han, D., Li, T., Yang, D., and Dong, A. (2016) Free-standing, ordered mesoporous few-layer graphene framework films derived from nanocrystal superlattices self-assembled at the solid- or liquid-air interface. *Chemistry of Materials*, **28**, 3823-3830.

12. Talapin, D. V., Lee, J. S., Kovalenko, M. V., and Shevchenko, E. V. (2010) Prospects of colloidal nanocrystals for electronic and optoelectronic applications. *Chemical Reviews*, **110**, 389-458.

13. Zhao, D., Huo, Q., Feng, J., Chmelka, B. F., and Stucky, G. D. (2014) Nonionic triblock and star diblock copolymer and oligomeric surfactant syntheses of highly ordered, hydrothermally stable, mesoporous silica structures. *Journal of the American Chemical Society*, **136**, 10546-10546.

14. Noh, M., Kwon, Y., Lee, H., Cho, J., Kim, Y., and Kim, M. G. (2005) Amorphous carbon-coated tin anode material for lithium secondary battery. *Chemistry of Materials*, **17**, 1926-1929.

15. Ng, S. H., Wang, J., Wexler, D., Chew, S. Y., and Liu, H. K. (2007) Amorphous carbon-coated silicon nanocomposites: A low-temperature synthesis via spray pyrolysis and their application as high-capacity anodes for lithium-ion batteries. *Journal of Physical Chemistry C*, **111**, 11131-11138.

16. Kim, H.-S., Kong, M., Kim, K., Kim, I.-J., and Gu, H.-B. (2007) Effect of carbon coating on $LiNi_{1/3}Mn_{1/3}Co_{1/3}O_2$ cathode material for lithium secondary batteries. *Journal of Power Sources*, **171**, 917-921.

17. Alakoski, E., Kiuru, M., Tiainen, V.-M., and Anttila, A. (2003) Adhesion and quality test for tetrahedral amorphous carbon coating process. *Diamond and Related Materials*, **12**, 2115-2118.

18. Cao, Q., Zhang, H. P., Wang, G. J., Xia, Q., Wu, Y. P., and Wu, H. Q. (2007) A novel carbon-coated $LiCoO_2$ as cathode material for lithium ion battery. *Electrochemistry Communications*, **9**, 1228-1232.

19. Das, N. S., Banerjee, D., and Chattopadhyay, K. K. (2011) Enhancement of electron field emission by carbon coating on vertically aligned Si nanowires. *Applied Surface Science*, **257**, 9649-9653.

20. Show, Y. (2007) Electrically conductive amorphous carbon coating on metal bipolar plates for PEFC. *Surface and Coatings Technology*, **202**, 1252-1255.

21. Liu, H., Wang, G., Wang, J., and Wexler, D. (2008) Magnetite/carbon core-shell nanorods as anode materials for lithium-ion batteries. *Electrochemistry Communications*, **10**, 1879-1882.

22. Perez-Huerta, A., and Cusack, M. (2009) Optimizing electron backscatter diffraction of carbonate biominerals-resin type and carbon coating. *Microscopy and Microanalysis*, **15**, 197-203.

23. Kim, B. G., Choi, Y., Lee, J. W., Lee, Y. W., Sohn, D. S., and Kim, G. M. (2000) Multi-layer coating of silicon carbide and pyrolytic carbon on UO_2 pellets by a combustion reaction. *Journal of Nuclear Materials*, **281**, 163-170.
24. Lopez-Honorato, E., Chiritescu, C., Xiao, P., Cahill, D. G., Marsh, G., and Abram, T. J. (2008) Thermal conductivity mapping of pyrolytic carbon and silicon carbide coatings on simulated fuel particles by time-domain thermoreflectance. *Journal of Nuclear Materials*, **378**, 35-39.
25. Kim, J., Kim, B., Lee, J.-G., Cho, J., and Park, B. (2005) Direct carbon-black coating on $LiCoO_2$ cathode using surfactant for high-density Li-ion cell. *Journal of Power Sources*, **139**, 289-294.
26. Ponrouch, A., Goni, A. R., Sougrati, M. T., Ati, M., Tarascon, J.-M., Nava-Avendano, J., and Palacin, M. R. (2013) A new room temperature and solvent free carbon coating procedure for battery electrode materials. *Energy & Environmental Science*, **6**, 3363-3371.
27. Rodriguez-Mirasol, J., Cordero, T., Radovic, L. R., and Rodriguez, J. J. (1998) Structural and textural properties of pyrolytic carbon formed within a microporous zeolite template. *Chemistry of Materials*, **10**, 550-558.
28. Lin, Y., Gao, M. X., Zhu, D., Liu, Y. F., and Pan, H. G. (2008) Effects of carbon coating and iron phosphides on the electrochemical properties of $LiFePO_4/C$. *Journal of Power Sources*, **184**, 444-448.
29. Klebanoff, L. E., Clift, W. M., Malinowski, M. E., Steinhaus, C., Grunow, P., and Bajt, S. (2002) Radiation-induced protective carbon coating for extreme ultraviolet optics. *Journal of Vacuum Science & Technology*, **B20**, 696-703.
30. Wang, J., Yang, J., Tang, Y., Li, R., Liang, G., Sham, T.-K., and Sun, X. (2013) Surface aging at olivine $LiFePO_4$: a direct visual observation of iron dissolution and the protection role of nano-carbon coating. *Journal of Materials Chemistry*, **A1**, 1579-1586.
31. Nilsson, L., Andersen, M., Balog, R., Laegsgaard, E., Hofmaan, P., Basenbacher, F., Hammer, B., Stensgaard, I., and Hornekaer, L. (2012) Graphene coatings: Probing the limits of the one atom thick protection layer. *ACS Nano*, **6**(11), 10258-10266.
32. Prasai, D., Tuberquia, J. C., Harl, R. R., Jennings, G. K., and Bolotin, K. I. (2012) Graphene: Corrosion-inhibiting coating. *ACS Nano*, **6**(2), 1102-1108.
33. Nishihara, H., Fukura, Y., Inde, K., Tsuji, K., Takeuchi, M., and Kyotani, T. (2008) Carbon-coated mesoporous silica with hydrophobicity and electrical conductivity. *Carbon*, **46**, 48-53.
34. Yoshida, K., Matsukawa, K., Imai, M., and Yano, T. (2009) Formation of carbon coating on SiC fiber for two-dimensional SiC_f/SiC composites by electrophoretic deposition. *Materials Science and Engineering, Part B*, **161**, 188-192.

35. Lee, H.-Y., and Lee, S.-M. (2004) Carbon-coated nano-Si dispersed oxides/graphite composites as anode material for lithium ion batteries. *Electrochemistry Communications*, **6**, 465-469.
36. Li, B., and Zhang, J. (2015) Polysiloxane/multiwalled carbon nanotubes nanocomposites and their applications as ultrastable, healable and superhydrophobic coatings. *Carbon*, **93**, 648-658.
37. Sevast'yanov, V. G., Kolesnikov, V. A., Desyatov, A. V., Kolesnikov, A. V. (2015) Conducting coatings based on carbon nanomaterials and SnO_2 on glass for photoconverters. *Glass and Ceramics*, **71**, 439-442.
38. Wang, H., Yang, Y., Liang, Y., Robinson, J. T., Li, Y., Jackson, A., Cui, Y., and Dail H. (2011) Graphene-wrapped sulfur particles as a rechargeable lithium-sulfur battery cathode material with high capacity and cycling stability. *Nano Letters*, **11**(7), 2644-2647.
39. Cushing, B. L., and Goodenough, J. B. (2002) Influence of carbon coating on the performance of a $LiMn_{0.5}Ni_{0.5}O_2$ cathode. *Solid State Sciences*, **4**, 1487-1493.
40. Chmiola, J., Largeot, C., Taberna, P. L., Simon, P., and Gogotsi, Y. (2010) Monolithic carbide-derived carbon films for micro-supercapacitors. *Science*, **328**, 480-483.
41. Chung, S.-H., and Manthiram, A. (2014) A polyethylene glycol-supported microporous carbon coating as a polysulfide trap for utilizing pure sulfur cathodes in lithium-sulfur batteries. *Advanced Materials*, **26**, 7352-7357.
42. Roldan, L., Santos, I., Armenise, S., Fraile, J. M., and Garcia-Bordeje, E. (2012) The formation of a hydrothermal carbon coating on graphite microfiber felts for using as structured acid catalyst. *Carbon*, **50**, 1363-1372.
43. Zhu, Z., Cheng, F., and Chen, J. (2013) Investigation of effects of carbon coating on the electrochemical performance of $Li_4Ti_5O_{12}$/C nanocomposites. *Journal of Materials Chemistry*, **A1**, 9484-9490.
44. Zhang, W.-M., Wu, X.-L., Hu, J.-S., Guo Y.-G., and Wan, L.-J. (2008) Carbon coated Fe_3O_4 nanospindles as a superior anode material for lithium-ion batteries. *Advanced Functional Materials*, **18**, 3941-3946.
45. Wang, H., Xu, Z., Kohandehghan, A., Li, Z., Cui, K., Tan, X., Stephenson, T. J., King'ondu, C. K., Holt, C. M. B., Olsen, B. C., Tak, J. K., Harfield, D., Anyia, A. O., and Mitlin, D. (2013) Interconnected carbon nanosheets derived from hemp for ultrafast supercapacitors with high energy. *ACS Nano*, **7**(6), 5131-5141.
46. Cano-Serrano, E., Blanco-Brieva, G., Campos-Martin, J. M., and Fierro, J. L. G. (2003) Acid-functionalized amorphous silica by chemical grafting-quantitative oxidation of thiol groups. *Langmuir*, **19**, 7621-7627.
47. Figueiredo, J. L., Pereira, M. F. R., Freitas, M. M. A., and Orfao, J. J. M. (1999) Modification of the surface chemistry of activated carbons. *Carbon*, **37**, 1379-1389.

48. Zawadzki, J., Wisniewski, M., Weber, J., Heintz, O., and Azambre, B. (2001) IR study of adsorption and decomposition of propan-2-ol on carbon and carbon-supported catalysts. *Carbon*, **39**, 187-192.
49. Zielke, U., Huttinger, K. J., and Hoffman, W. P. (1996) Surface-oxidized carbon fibers: I. Surface structure and chemistry. *Carbon*, **34**, 983-998.

4

Self-Healing Anti-Corrosion Coatings Embedded with Graphene and Graphene Oxide based Containers Encapsulating Corrosion Inhibitor in Polyelectrolyte Complexes

4.1 Introduction

The destruction of metal structures by corrosion is one of the major challenges faced by the petrochemical, automobile and aerospace industries owing to the significant economic losses associated with the remediation and maintenance processes [1]. An effective strategy commonly employed to prevent corrosion is the application of polymeric coatings, which provide a passive protection by hindering the moisture and other corrosive ions from reaching the metal surface. The continuous exposure of these coatings to severe environmental conditions creates cracks or micro-pores leading to the diffusion of the corrosive species to the metal surface, thus, slowly diminishing the passive protection nature of the coatings [2,3]. The direct incorporation of the corrosion inhibitor in the coatings though provides an active protection, however, it also leads to the environmental as well as health challenges due to the leaching of the inhibitors (e.g., Cr(VI) species) from the coating matrix [4-7].

Self-healing coatings, fabricated by the incorporation of the corrosion inhibitor containers in the coating matrix, are a beneficial alternative to the inhibitor loaded conventional coatings [8-15]. The containers can isolate the inhibitor from the coating matrix, thus, avoiding the negative impact on the integrity of the coating. The containers are sensitive to the external (e.g., mechanical damage) and internal (e.g., pH changes) corrosion triggers, and, therefore, as the environment of the coating changes, the smart containers swiftly respond by releasing the inhibitor (healing agent) onto the metal surface to eliminate/delay the corrosion process. Also, the containers prevent the unwanted leaching of the corrosion inhibitor from the coatings, thus, maintaining their environmental safety. Several types of containers have been developed based on the encapsulation techniques [6-

Gisha Elizabeth Luckachan and Vikas Mittal, Khalifa University of Science and Technology, Abu Dhabi, UAE
© *2021 Central West Publishing, Australia*

8,8,16-21]. Among these, the containers employing the polyelectrolyte shells represent the most promising approach owing to the stimulation by pH shift to release the corrosion inhibitor [18-21]. These containers are obtained by the layer-by-layer (lbl) addition of the polyelectrolytes and corrosion inhibitor on the surface of the charged metal oxide nanoparticles. As the corrosion process results in a decrease of pH at the anodic areas and an increase of pH at the cathodic areas [22], it results in the high efficiency of the containers triggered by the pH changes. Correspondingly, the release of the corrosion inhibitor from the shells is also hindered as soon as the healing process brings the pH of the corrosion site to neutral. Shchukin et al. [8,11,16,18,22] reported for the first time the use of the polyelectrolyte shells-based containers for fabricating the active anti-corrosion coatings. Silica nanoparticles and halloysite nanotubes were chosen as the supporting hosts for the encapsulation of benzotriazole between the oppositely charged polyelectrolyte layers. The resulting nano-containers were coated on the aluminum alloy using the ZrO_2/SiO_2 sol-gel coatings.

Graphene (Gr), atomically thick two-dimensional layers of sp^2 carbon atoms bonded together in a honeycomb lattice, has been recently reported as an ideal corrosion inhibition agent due to unique chemical and thermal resistance, mechanical strength, electrical and thermal conductivities, etc. [23-31]. Graphene prevents the penetration of moisture and other corrosive species like oxygen and chlorides due to the barrier effect owing to its high aspect ratio and platelet structure. Graphene also results in electron depletion in direct contact with the metal, thus, leading to the passivation of the metal/metal oxide surface through the establishment of a Schottky barrier, thereby, protecting the metal against corrosion by an active-passive approach [23,32]. In this respect, several studies have reported the corrosion inhibitive nature of graphene in the polymer composites based protective coatings [23-33].

To the best of our knowledge, the use of graphene as a host for the nano-container synthesis for effective corrosion protection has not been studied. Recently, Cui et al. [34] reported the fabrication of multi-walled carbon nanotubes (MWCNTs)-embedded microcapsules by the stepwise deposition of the polyelectrolytes and corrosion inhibitor using the lbl self-assembly technique based on electrostatic interaction. The resulting microcapsules exhibited optimal electrochemical behavior for application in biosensing and catalysis. To explore the development of the nano-containers based on graphene, the lbl addition of the polyelectrolytes and corrosion inhibitor on the surface of graphene particles was explored in this study. The obtained nano-containers were subsequently embedded in the polymeric coatings to study the anti-corrosion behavior. Such functional containers can be envisaged to provide an active corrosion protection owing to the controlled release of the corrosion inhibitor as well as a passive protection by enhancing the diffusion path of the corrosive agents through the coating. The encapsulation efficiency as well as release kinetics of the different corrosion inhibitors from the graphene-based containers were also studied. For comparison, graphene oxide (GO)-based containers were also fabricated and studied for their performance.

4.2 Experimental

4.2.1 Materials

Carbon steel coupons of size 2 cm x 5 cm were purchased locally and used as coating substrate. The composition of carbon steel in weight percentage was C (0.125), Mn (0.519), Si (0.016), P (0.014), S (0.005), Al (0.034) and Fe (99.287). Vinyl acrylate, used as the polymer matrix, and acrylic thinner were purchased from National Protective Coatings, UAE. Poly(sodium 4-styrenesulfonate) (PSS) and polyallylamine hydrochloride (PAH), the polyelectrolytes used for the nano-container synthesis, were procured from Sigma Aldrich. Benzotriazole (BTA, 99%) and 2-mercaptobenzothiazole (MBT), used as corrosion inhibitors, were also supplied by Sigma Aldrich. Hydrochloric acid used for the pickling of the coating substrate (37%) was purchased from Merck. Sandpapers of 60, 150 and 180 grit sizes were purchased from ACE Hardware, UAE. Two-pack epoxy hardener and base were locally procured. Single layer graphene nanoplatelets were supplied by Angstron Materials, USA. Graphene oxide was synthesized as per the procedure reported earlier [35].

4.2.2 Preparation of Gr and GO Based Containers

0.5 g graphene was dispersed in 25 mL deionized water using probe sonication for 5 min, followed by 20 min ultrasonication and 1 h magnetic stirring. 50 mg positively charged PAH was dissolved in 5 mL deionized water and was added as the first layer. The mixture was stirred for 15 min, followed by the removal of excess PAH by centrifugation and repeated washing with water. A second layer of the negatively charged PSS was achieved by adding the mixture of 50 mg PSS in 5 mL deionized water. The mixture was stirred again for 15 min, and the excess PSS was similarly removed by centrifugation and washing. For the third layer, the solution prepared by dissolving 200 mg BT in 10 mL deionized water (pH 2) was added to the above mixture and stirred for 20 min. The second and third steps were repeated twice, followed by the fabrication of the final layers of PSS and PAH. Thus, the final lbl structure formed around the graphene nanoplatelets was G/PAH/(PSS/BT)$_2$/PSS/PAH. Scheme 4.1 presents the schematic representation of the container depicting the polyelectrolyte layers and encapsulated corrosion inhibitor on the surface of graphene platelets.

Similar steps were repeated using MBT as corrosion inhibitor to achieve G/(PAH/MBT)$_2$/PAH/PSS nano-containers. GO based nano-containers encapsulating BT and MBT were also prepared using the same steps. The final structure of the GO-based containers is represented as GO/PAH/(PSS/BT)$_2$/PSS/PAH and GO/(PAH/MBT)$_2$/PAH/PSS.

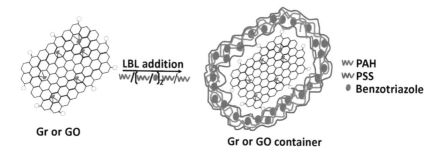

Scheme 4.1 Schematic representation of the addition of polyelectrolyte and corrosion inhibitor onto the surface of graphene or graphene oxide platelets.

4.2.3 Zeta Potential

Zeta potential measurements were performed at each step during the nano-container synthesis to check the potential of each layer in order to ensure a strong bonding with the surface species. The instrument used for this purpose was ZETA PALS with the mobility range from 10^{-11} to 10^{-7} m^2/V*s.

4.2.4 Preparation of Composite Coatings

Commercially available solvent-based vinyl acrylate was used as the matrix for developing the coatings incorporating the nano-containers. Specifically, the coating formulations were prepared by mixing 0.2 g nano-container powders with 15 g vinyl acrylate in 100 mL solvent under continuous magnetic stirring for 1 h. The consistency of the mixture was maintained by adding 2 mL acrylic thinner. The obtained formulation was coated on carbon steel coupons using dip coating. An immersion time of 60 s and a withdrawal speed of 100 mm/min were used for the first coating. After 1-2 minutes of drying under ambient conditions, a second coating with an immersion time of 30 s was applied. The coated substrates were dried in an oven at 110 °C for 2 h, followed by overnight drying at 60 °C. The edges of the coated substrates were sealed by applying an epoxy resin prepared by mixing the base and hardener components in 4:1 weight ratio. Five specimens were prepared for each coating formulation.

4.2.5 Fourier-transform Infrared Spectroscopy (FTIR)

The Fourier-transform infrared spectroscopy of the coatings was performed using a Bruker VERTEX 70 FTIR spectrometer attached with a transition accessory. Gr and GO pellets were also prepared by mixing with KBr for IR analysis. IR acquisition was achieved by collecting 120 scans at a resolution of 4 cm^{-1} in the frequency range of 370 cm^{-1} to 4000 cm^{-1} using OPUS software.

4.2.6 Immersion Tests

The coated coupons were immersed in 0.35 M sodium chloride solution at room temperature, and the corrosion was qualitatively monitored by periodically checking the formation of the corrosion products or rust. With the help of optical microscopy (Digital Microscope KH-7700, Hirox Co. Ltd, USA), the coated substrates were carefully inspected for any signs of corrosion initiation and local delamination.

4.2.7 UV Weathering Analysis

UV weathering analysis of the surface of the coatings was carried out in a UV light accelerated ageing cabinet (BGD 856 from Buiged Laboratory Instruments, Guangzhou), equipped with 4 fluorescent UV 6 lamps (UVA-340) oriented horizontally on both sides in accordance with ASTM D4329. The samples were exposed to 4 cycles of alternate UV and water condensation processes. UV weathering accelerated conditions for each cycle were set as follows: 0.67 Watt/m² of UV irradiance for 8 h at 50 °C, followed by water condensation for 4 h at 50 °C. After completion of 4 cycles, the electrochemical impedance spectroscopy (EIS) and Tafel measurements were carried out.

4.2.8 Electrochemical Analysis

A three-electrode cell (250 mL volume) with a platinum gauze counter electrode and a saturated calomel reference electrode (SCE) with bridge tube was used to perform the electrochemical tests on the flat coated coupons. The exposed working electrode area was 3.6 cm², and the coated coupons were kept in place with the help of an O ring and adjustable brass disc. The tests were carried out by connecting the corrosion cell to the BioLogic VMP-300 multi-potentiostat (controlled by a computer running EC-Lab 10.40 software) employing 0.35 M NaCl solution at room temperature. The ultra-low current cable connected to the potentiostat was used for the accurate measurement of the current. This option includes the current ranges from 100 nA down to 100 pA with additional gains extending the current ranges to 10 pA and 1 pA. The resolution on the lowest range is 76 aA. The open circuit potential (OCP) was measured for 5 min in order to allow the potential to stabilize before the electrochemical impedance and potentiodynamic polarization analyses. The impedance measurements were performed at an amplitude of 15 mV over the frequency range from 200 kHz to 10 MHz. The same software was also used to simulate the impedance behavior of the samples. Tafel plots were measured by polarizing the working electrode from an initial potential of -250 mV up to a final potential of +250 mV as a function of open circuit potential. A scan rate of 0.1667 mV/s was used for the polarization sweep. For reproducibility of the measurements, each set of experiments were repeated three times on the newly coated samples.

4.3 Results and Discussion

In order to study the effect of carbon as a host matrix for the encapsulation of corrosion inhibitor, two different types of containers were prepared by layer by layer addition of polyelectrolytes and corrosion

Self-Healing Anti-Corrosion Coatings

inhibitor onto the surface of graphene (Gr) and graphene oxide (GO). Two different corrosion inhibitors, benzotriazole and 2-mercaptobenzotriazole, were used to impregnate the polyelectrolyte shell on the surface of the Gr and GO containers. The charging of the containers with the adsorption of polyelectrolyte and corrosion inhibitor layers at each step of addition onto the surface of Gr and GO was monitored by electrophoretic measurements, as shown in Figure 4.1. A negative zeta potential was observed for Gr and GO at the beginning

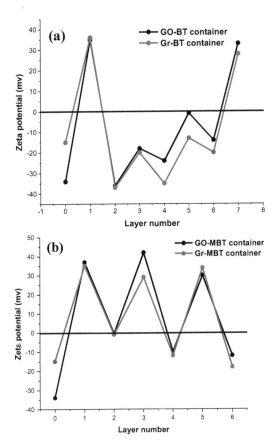

Figure 4.1 Changes in the surface charges of Gr and GO with the addition of different layers of polyelectrolytes and corrosion inhibitors. (a) 0: Gr or GO; 1: (Gr or GO)/PAH; 2: (Gr or GO)/PAH/PSS; 3: (Gr or GO)/PAH/PSS/BT; 4: (Gr or GO)/PAH/PSS/BT/PSS; 5: (Gr or GO)/PAH/PSS/BT/PSS/BT; 6: (Gr or GO)/PAH/(PSS/BT)2/PSS and 7: (Gr or GO)/PAH/(PSS/BT)2/PSS/PAH. (b) 0: Gr or GO; 1: (Gr or GO)/PAH; 2: (Gr or GO)/PAH/MBT; 3: (Gr or GO)/PAH/MBT/PAH; 4: (Gr or GO)/PAH/MBT/PAH/MBT; 5: (Gr or GO)/(PAH/MBT)2/PAH and 6: (Gr or GO)/ (PAH/MBT)2/PAH/PSS.

of container preparation which indicated that a significant amount of polar functional groups was present on the surface of graphene and graphene oxide, leading to the negative surface potential. The addition of first layer of PAH drastically changed the surface charges of Gr and GO to positive side, as shown in Figure 4.1a. The addition of the negative polyelectrolyte PSS as second layer shifted the charges to negative side as well. The changes in surface the charges were less pronounced after the addition of benzotriazole as third layer. This is due to the small size of the benzotriazole molecules, which could neutralize only the excess negative charges on the container surface. The multi-charged long chain PAH and PSS polymers have strong electrostatic forces, and, therefore, these can be adsorbed in sufficient quantities to completely recharge the surface [19,20]. Two layers of PSS/benzotriazole were incorporated in the polyelectrolyte shell to ensure the encapsulation of a sufficient amount of corrosion inhibitor in the final container. The last layer of PAH created a net positive charge on the surface of Gr/PAH/(PSS/BT)2/PSS/PAH and GO/PAH/(PSS/BT)2/PSS/PAH containers. Gr/(PAH/MBT)2/PAH/PSS and GO/(PAH/MBT)2/PAH/PSS containers were also prepared by changing the corrosion inhibitor with 2-mercaptobenzotriazole and the charging of containers after the addition of each layer is shown in Figure 4.1b. The order of layers in the MBT encapsulated container was changed based on the charging of the surface after the addition of polyelectrolytes and MBT layers. MBT was incorporated in between the positively charged PAH layers. Two layers of PAH/MBT were incorporated to ensure the encapsulation of a sufficient amount of corrosion inhibitor inside the polyelectrolyte shell. PSS was added as the final layer (6th) onto the surface of the containers. Therefore, the final layer of the BT incorporated containers was positively charged PAH, whereas for MBT incorporated containers, it was negatively charged PSS. The number of layers in the polyelectrolyte shell was seven for BT incorporated containers, whereas it was six for MBT incorporated containers.

The sol-gel process is the commonly used technique for the formulation of container coatings on metal surfaces. However, the sol-gel coatings experience several drawbacks like highly porous structure, low mechanical integrity and long processing duration. Sintering at high temperatures also limits the application of the sol-gel coatings on temperature sensitive substrates and devices. In addition, the preparation of thick films leads to cracking problems as well [36-39]. These drawbacks limit the application of the sol-gel coatings for the long-term protection in industrial sectors. Incorporation of containers in a primer coating would be a beneficial method to achieve a long-term active protection of the metal substrates. Therefore, the coating formulations of the carbon containers were prepared by choosing solvent based vinyl acrylate primer, commonly used as a first layer in the full coating system. The carbon containers were dispersed in vinyl acrylate (VA) matrix and subsequently dip-coated onto carbon steel surface. The optical images of the coatings presented in Figures 4.2 b and c demonstrated the uniform distribution of graphene containers in the coatings (VA/Gr-BT and VA/Gr-MBT). However, the agglomeration of the graphene oxide particles was apparent in the VA/GO-BT and VA/GO-MBT coatings, as shown in Figures 4.2 d and e. Figure 4.2a shows the optical image of the pure VA

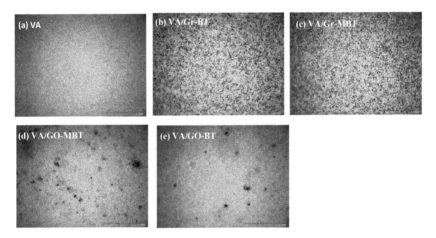

Figure 4.2 Optical images of vinyl acrylate and carbon containers incorporated vinyl acrylate coatings on mild carbon steel.

coating. The outer layers of PAH and PSS respectively on the BT and MBT encapsulated containers were compatible with the vinyl acrylate matrix, as confirmed by the uniform distribution of Gr-BT and Gr-MBT containers in the coating matrix (Figures 4.2 b and c). Therefore, the agglomerated GO-BT and GO-MBT containers in the vinyl acrylate matrix could be explained based on the hydrophilic nature of GO as compared with Gr (Figures 4.2 d and e). The FTIR spectra of Gr and GO in Figure 4.3 clearly revealed the presence of hydrophilic functional groups on the graphene oxide surface. The main absorption band at 3340 cm^{-1} was assigned to the O-H stretching vibrations of the carboxyl and hydroxyl functional groups. The absorption peaks at

1730 cm⁻¹ and 1624 cm⁻¹ could be assigned to C=O stretching of the carboxylic and/or carbonyl functional groups. The absorption peak at about 1044 cm⁻¹ was assigned to the C–O stretching vibrations [40,41]. Such characteristic vibrational peaks were not significantly visible in the IR spectrum of graphene, however, the presence of hydrophilic functional groups could be confirmed from the zeta potential measurements which indicated a negatively charged surface of graphene. The hydrophilic functional groups in GO led to a strong interaction among the different layers, thus, enhancing the chances of agglomeration during the assembly of different layers and coating deposition on the carbon steel surface. The thickness of the developed coatings was determined to be 190±20 μm.

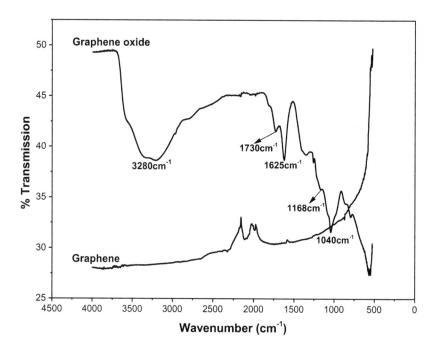

Figure 4.3 The FTIR spectra of graphene and graphene oxide.

The graphene and graphene oxide based anti-corrosion coatings are generally prepared for the active and passive protection of metal substrates. A majority of the research studies have reported that the carbon materials enhance the corrosion protection of the polymeric coatings by increasing the tortuous path of water and corrosive species through the coatings [23-33]. The role of Gr and GO as carriers of

Self-Healing Anti-Corrosion Coatings

corrosion inhibitors has been seldom reported in the literature. In the present work, the anti-corrosion nature of vinyl acrylate coatings embedded with corrosion inhibitor encapsulated Gr and GO containers was analyzed in 0.35 M sodium chloride solution. Figure 4.4 presents the digital images of the carbon coatings along with pure vinyl acrylate coating after four weeks of immersion in sodium chloride solution. The coatings were observed to be largely intact, and the central

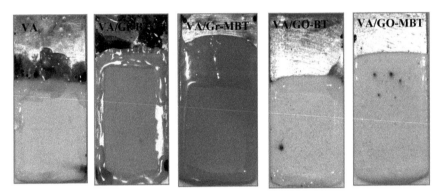

Figure 4.4 Digital images of the coatings after 4 weeks of immersion in 0.35 M sodium chloride solution.

area of the coated coupons remained without corrosion. The corrosion spots appeared mostly on the VA/GO-BT coating at the edges near the epoxy overcoat. A few corrosion spots were noticed near the center of the VA/GO-MBT coating as well. In Gr-BT coating, the corrosion started at the edges of the carbon steel beneath the epoxy covering. The propagation of defects from the edges towards the central coating area could be clearly observed. However, such a defect propagation was not observed in the VA/Gr-BT coating. As vinyl acrylate as such provides a barrier, it is difficult to compare the properties of the different coatings simply by the immersion test. Therefore, EIS in 0.35 M sodium chloride solution was employed to quantify the corrosion protection behavior of different coatings. As observed from Figure 4.5, the overall impedance and corrosion resistance of the vinyl acrylate coating were reduced on the incorporation of carbon containers. The low frequency impedance modulus ($Z_{0.01Hz}$) is commonly used to obtain a rough estimate of the coating performance. The incorporation of carbon containers decreased the low frequency impedance of the vinyl acrylate coating from 10^6 Ωcm^2 to 10^4-10^5 Ωcm^2. Prolonged immersion in the sodium chloride solution gradually decreased the low frequency impedance of the vinyl acrylate coating to 10^4 Ωcm^2 after four weeks. The carbon coatings exhibited a

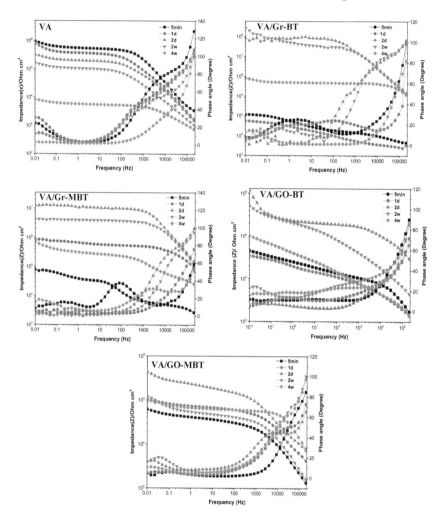

Figure 4.5 Bode and phase plots of the coatings as a function of immersion duration in 0.35 M sodium chloride solution.

different behavior as compared to the pure vinyl acrylate coating, where the low frequency impedance increased after immersion for 2 d, followed by a slow reduction to a value higher than the initial low frequency impedance at the end of 4 weeks. The observed impedance behavior could be attributed to the active protection of the carbon coatings. The corrosion processes on the metal surface triggered the release of benzotriazole from the carbon containers, which hindered the corrosive species from reaching the metal surface by forming a stable layer at the metal/coating interface, thus, enhancing the overall resistance of the coating. For Gr-BT based coatings, the low frequency impedance increased to 10^8-10^9 Ωcm^2 after 2 d and remained in the same range over 2 weeks. The initial low frequency impedance of the carbon-based coatings was in the range 10^4-10^5 Ωcm^2. The Gr-BT and Gr-MBT coatings exhibited a value of 10^6 Ωcm^2, whereas the GO-BT and GO-MBT coatings showed a value close to 10^5 Ωcm^2 after 4 weeks. It indicated that though the carbon coatings provided an active protection, the graphene-based coatings were more efficient than the graphene oxide-based coatings. It could be attributed to the availability of benzotriazole at the required areas to hinder the corrosion processes owing to the uniform distribution of the graphene-based containers in the vinyl acrylate matrix, as observed in the optical images in Figure 4.4.

The carbon containers in the vinyl acrylate coating play two roles. These maintain the coating integrity by avoiding a direct contact between the inhibitor and polymer matrix, along with providing a controlled delivery of inhibitor at the defect sites to ensure the active protection. It can be analyzed from the coating resistance and charge transfer resistance parameters. For this, the coatings were fitted with an equivalent circuit shown in Figure 4.6. The symbol R_s represents the solution resistance of the bulk electrolyte between the reference

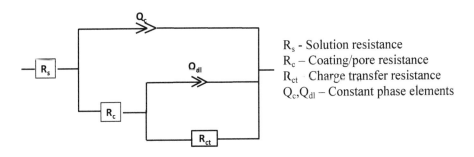

Figure 4.6 Equivalent circuit used to fit the EIS plots of the vinyl acrylate and carbon container incorporated vinyl acrylate coatings.

electrode and working electrode, R_c is the coating resistance, R_{ct} is the charge transfer resistance, and Q_c and Q_{dl} are the constant phase elements (CPE) used instead of the pure capacitances C_c and C_{dl}, respectively [40]. The coating resistance is related to the resistance of the electrolytes in the pores or cracks. Therefore, the monitoring of the coating resistance is expected to provide information on the porosity and degree of degradation of the vinyl acrylate coatings in the presence of carbon containers. Table 4.1 shows the time dependent evolution of the coating resistance (R_c) of the pure vinyl acrylate and VA/carbon container coatings. The incorporation of containers led to

Table 4.1 EIS parameters of the coatings after the fitting analysis

		R_c (Ω cm^2)	Q_c			R_{ct} (Ω cm^2)	Q_{dl}	
			Yo (Ω^{-1} cm^2)	n_1			Yo (Ω^{-1} cm^2)	n_2
VA	5 min	5.205e5	2.57e-9	0.899		9.420e6	9.08e-6	0.575
	1 d	3.667e5	17.89e-9	0.890		8.620e5	15.61e-6	0.814
	2 d	2.017e5	37.11e-9	0.898		1.988e5	41.65e-6	0.779
	2 w	1.062e5	79.09e-9	0.876		9.317e4	84.94e-6	0.567
	4 w	5.533e3	11.20e-6	0.894		2.694e3	0.77e-3	0.575
VA/Gr-BT	5 min	1.588e4	3.42e-6	0.861		1.121e4	0.11e-6	0.879
	1 d	1.950e4	8.83e-6	0.887		7.452e4	12.9e-6	0.883
	2 d	3.059e6	16.67e-6	0.890		9.026e6	13.73e-6	0.801
	2 w	4.056e6	24.36e-6	0.877		9.912e6	21.7e-6	0.793
	4 w	4.721e5	66.57e-6	0.880		9.095e5	60.5e-6	0.779
VA/Gr-MBT	5 min	2.945e4	1.34e-6	0.897		5.421e4	1.17e-6	0.900
	1 d	4.792e5	10.27e-6	0.892		3.363e4	4.60e-6	0.864
	2 d	9.769e6	24.10e-6	0.880		3.782e6	16.16e-6	0.843
	2 w	7.279e6	30.97e-6	0.877		4.325e6	30.86e-6	0.742
	4 w	3.879e5	62.61e-6	0.870		6.937e5	74.23e-6	0.754
VA/GO-BT	5 min	1.411e4	0.57e-9	0.869		2.715e4	0.25e-6	0.899
	1 d	1.331e4	2.12e-9	0.876		2.459e4	8.93e-6	0.853
	2 d	2.632e5	25.73e-9	0.886		5.798e5	15.97e-6	0.942
	2 w	2.852e5	96.98e-9	0.877		6.884e5	30.64e-6	0.733
	4 w	1.056e4	1.58e-8	0.880		8.666e4	54.67e-6	0.735
VA/GO-MBT	5 min	5.879e4	0.55e-9	0.879		4.762e4	10.11e-6	0.589
	1 d	6.538e4	37.08e-9	0.862		5.884e4	38.19e-6	0.814
	2 ds	1.821e5	47.85e-9	0.863		6.223e5	41.36e-6	0.767
	2 w	2.752e5	60.59e-9	0.874		1.288e5	44.26e-6	0.656
	4 w	3.214e4	8.64e-8	0.860		4.770e4	69.4e-6	0.780

the creation of irregularities on the coating surface, resulting in a decreased coating resistance of the VA/container coatings as compared to the pure VA coating. The continuous immersion decreased the resistance of the pure vinyl acrylate coating, whereas the carbon containers-based coatings showed an increasing tendency, which maintained over 2 weeks. These changes indicated that the VA coating was more prone to water molecules on enhancing the duration of immersion. With respect to the carbon container coatings, though these were more prone to the water molecules at the beginning of immersion due to the irregularities at the surface, however, the direct ingress of the electrolyte through the coatings was not possible. It can be further explained based on the CPE values provided in Table 4.1. The constant phase element permits the simulation of the phenomenon that deviates from a pure capacitive behavior [42-44]. The magnitude of CPE and capacitance are related to each other as per the equation:

$$C = Y_0 (w_{max})^{n-1}$$

where, Y_0 is the magnitude of CPE, w_{max} is the frequency at which the imaginary impedance reaches the maximum for the respective time constant, and n is the exponential term of CPE which can vary between 1 for a pure capacitor and 0 for a pure resistor [42,43]. As per the equation, CPE would be very similar to a pure capacitor when the constant n reaches close to 1, therefore, the constant Y_0 could follow the same trend as the capacitance [45]. The capacitance of the coating (C_c) is proportional to its dielectric constant and can be, therefore, attributed to the amount of water absorbed by the coating [46]. Therefore, by evaluating the changes in Y_0 of Q_c, the tendency of the coatings towards water intake can be monitored. A continuous and rapid increment in the Y_0 values was observed for the pure VA coating. The carbon containers-based coatings also showed an increase in the Y_0 values, however, the rate of increase was significantly slower than the pure VA coating, especially for the VA/Gr-BT and VA/Gr-MBT coatings. It should be noted that the initial value of Y_0 of the carbon containers-based coatings was higher than the pure VA coating. The observed findings can be explained based on the fact that the irregularities created on the VA coating by the incorporation of carbon containers led to an increased water intake during the initial hours of immersion. However, the diffused water molecules did not reach the metal surface as fast as the pure VA coating due to the increased tortuous path for the diffusion of the water molecules through the coating resulting due to the presence of the carbon containers. A schematic representation of the tortuous path of the water molecules through carbon containers-based coatings has been presented in Scheme 4.2a. Table 4.1 also shows the evolution of the charge transfer resistance (R_{ct}) of the samples, which helps to understand the behavior of containers in delivering the inhibitor at the defect site for an active corrosion protection. The charge transfer resistance is related to the extent of corrosion processes. For the vinyl acrylate coating, the resistance decreased with the time of immersion similar to a barrier coating, whereas an increase in R_{ct} was observed after 2 d for the carbon containers-based coatings. The observed behavior could be attributed to the self-healing nature of the carbon-based coatings.

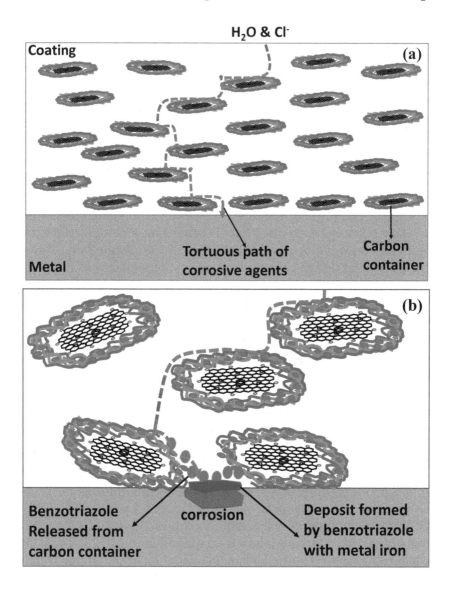

Scheme 4.2 A schematic representation of the (a) distribution of carbon containers in the VA coating on the metal surface and tge (b) release of corrosion inhibitor from the carbon containers during the immersion in 0.35 M sodium chloride solution.

In addition, the increased R_{ct} values were retained over 2 weeks of immersion duration, thus, demonstrating the effective long-term protection of the carbon container incorporated coatings. The initial low R_{ct} values compared with the pure vinyl acrylate coating resulted from the ingress of the electrolytes down to the metal surface which initiated the corrosion processes. The local corrosion activity triggered the release of the corrosion inhibitor from the containers based on the pH changes at the defect sites. Both benzotriazole and mercaptobenzotriazole have been reported as effective corrosion inhibitors for carbon steel substrates in chloride solutions. Scheme 4.2b represents the release of the corrosion inhibitor from the carbon containers during the initiation of the corrosion processes on the metal surface. As per the scheme, the corrosion inhibitor released from the containers inhibited the corrosion process by forming a passive product on the metal surface which prevented the electrolytes from reaching the metal surface for further corrosion reactions. The deposition of passive layers on the metal surface could be explained further from the deposited layer capacitance (C_{dl}):

$$C_{dl} = \varepsilon\varepsilon_o(At^{-1})$$

where, ε and ε_o represent the dielectric constant of deposition and permittivity of free space (8.9 x 10^{-14} F cm^{-1}), respectively; A is the area of the exposed metal surface; and t is the deposited layer thickness [46,47]. As Q_{dl} was used instead of pure capacitance, any changes in Y_0 of Q_{dl} could be related to the nature of the deposited layers on the metal surface. Compared with the pure VA coating, the Y_0 values of the carbon containers-based coatings were noted to decrease, especially after 2 weeks of immersion. Thus, it could be suggested that the area of the metal surface exposed for the corrosion process decreased. The Y_0 values were also related to the thickness of the deposited layers, which increased with the time of immersion in sodium chloride solution [48,49]. Such a deposition of layers on the metal surface can be attributed to an increased n_2 value. The exponential term of the constant phase element 'n' also measures the surface inhomogeneity; the lower is its value, the higher is the surface roughening of the metal/alloy and vice versa [50]. Though the changes in the Y_0 values were insignificant, the increased R_{ct} and n_2 values for the carbon containers-based coatings compared with the pure VA coating confirmed the deposition of the protective layers on the metal surface owing to the reaction of the corrosion inhibitor released from carbon containers with metal ions. The changes in both R_c and R_{ct} values indicated the self-healing behavior of coatings which was more prominent in the graphene-based coatings than the graphene oxide-based coatings. It might be attributed to the availability of the containers near the defect areas due to the uniform distribution of the Gr based containers in the coating matrix as compared with the GO based containers.

After 4 weeks of immersion, the coatings were subjected to potentiodynamic measurements in 0.35 M NaCl solution, and the corresponding Tafel plots and Tafel data are presented in Figure 4.7 and Table 4.2. respectively. The corrosion potential of the vinyl acrylate

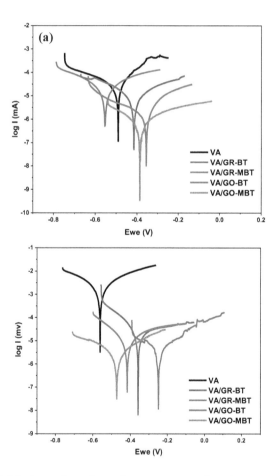

Figure 4.7 Tafel plots of vinyl acrylate coating and carbon containers incorporated vinyl acrylate coatings (a) before and (b) after 4-week immersion in 0.35 M NaCl solution.

Table 4.2 Tafel data of pure vinyl acrylate and vinyl acrylate mixed with carbon containers immersed in 0.35 M NaCl solution for 4 weeks

	Before immersion		After immersion	
	E_{corr} (mV)	I_{corr} (µA)	E_{corr} (mV)	I_{corr} (µA)
VA	-491	0.563	-562	1.021
VA/Gr-BT	-415	0.018	-358	0.013
VA/Gr-MBT	-553	0.107	-248	0.007
VA/GO-BT	-355	0.102	-409	0.009
VA/GO-MBT	-386	0.049	-472	0.010

coating (-562 mV) was close to the corrosion potential of carbon steel (-600 mV) after 4 weeks of immersion in 0.35 M NaCl solution. The carbon containers-based coatings exhibited the corrosion potential towards the positive side of the pure vinyl acrylate coating. After immersion, E_{corr} of VA/Gr-BT moved from -415 mV to -358 mV, whereas it moved from -553 mV to -248 mV for the VA/Gr-MBT coatings. However, the corrosion potential of the VA/GO-BT and VA/GO-MBT coatings shifted towards the cathodic side as compared to the initial values, thus, exhibiting a low degree of protection as compared to the graphene-based coatings. However, E_{corr} of the GO based coatings was in the positive side as compared to E_{corr} of the pure vinyl acrylate coating, which indicated the ability of GO based coatings in providing protection. The corrosion current of the VA/GO-BT and VA/GO-MBT coatings showed a significant reduction from 0.102 µA to 0.009 µA and 0.049 µA to 0.010 µA, respectively. The corrosion current of graphene-based coatings also showed a reduction from 0.018 µA to 0.013 µA and 0.107 µA to 0.007 µA for VA/Gr-BT and VA/Gr-MBT coatings, respectively. The changes in E_{corr} and I_{corr} confirmed the higher corrosion protection ability of the graphene-based coatings than the graphene oxide-based coatings. The overall increase in the corrosion protection ability of VA coating after the incorporation of the carbon containers could be further confirmed from the anodic and cathodic currents which decreased significantly for the carbon containers-based coatings as compared with the pure vinyl acrylate coating (Figure 4.7).

The carbon materials are widely used as stabilizers in the polymer products for absorbing or screening out the damaging UV light and transforming the energy into heat for harmless dissipation through the product. Therefore, the carbon containers-based coatings were subjected to UV weathering analysis to study their ability in providing the UV protection as well. The pure vinyl acrylate coating and carbon containers-based coatings were exposed to four cycles of alternate UV and water condensation. The digital images, presented in Figure 4.8, underline the difficulty in analyzing the performance of the coatings qualitatively, as all the coatings remained intact during

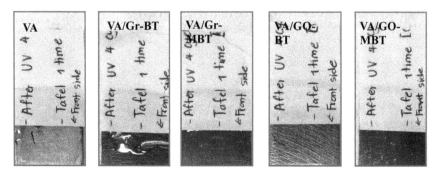

Figure 4.8 Digital images of the vinyl acrylate coating and carbon containers-based coatings after four cycles of alternate UV exposure and water condensation.

the UV exposure. Therefore, a quantitative measurement was conducted by EIS and Tafel analyses in 0.35 M NaCl solution after the exposure to four cycles of alternate UV and water condensation. The Bode plots of the pure vinyl acrylate presented in Figure 4.9 indicated that the UV exposure decreased the low frequency impedance of the vinyl acrylate coating from 10^6 Ωcm^2 to 10^4 Ωcm^2, signifying that the matrix was not UV protective. Both Gr and GO container coatings showed an increased low frequency impedance after the four cycles of UV weathering exposure as compared with the pure vinyl acrylate coating. It was obvious from the R_c and R_{ct} data provided in Table 4.3 where both coating resistance and charge transfer resistance of the carbon containers-based coatings increased as compared with the pure vinyl acrylate coating after UV exposure. The observed phenomenon might be attributed to the combined action of the UV protective behavior of the carbon materials as well as the inhibitive action of the corrosion inhibitor released from the carbon containers during the UV weathering conditions. Further, the graphene containers-based coatings provided a slightly enhanced protection as compared with the graphene oxide container embedded coatings due to the uniform dispersion and distribution of the containers in the vinyl acrylate matrix. The Tafel plots of the samples measured after 4 cycles of UV exposure and water condensation are presented in Figure 4.10. After 4 cycles of UV weathering analysis, the carbon containers-based coatings exhibited the E_{corr} values at the positive side of the pure vinyl

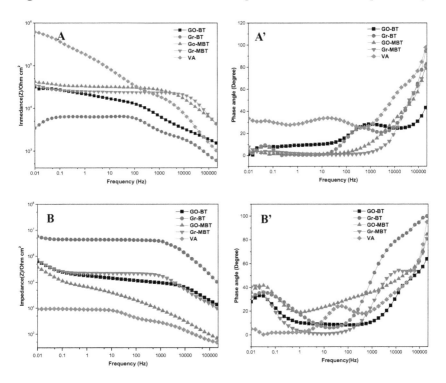

Figure 4.9 Bode and phase plots of the coatings (A & A') before and (B & B') after 4 cycles of UV weathering analysis.

Table 4.3 R_c, R_{ct}, E_{corr} and I_{corr} data of the coatings after 4 cycles of UV exposure and water condensation

		VA	VA/Gr-BT	VA/Gr-MBT	VA/GO-BT	VA/GO-MBT
R_c	Initial	5.95E6	2542	23432	11554	31566
	After 4 cycles	4518	427000	214444	185526	106018
R_{ct}	Initial	7.695E7	3828	10561	7690	11404
	After 4 cycles	47350	1.157E6	1.028E6	954034	748487
E_{corr}	After 4 cycles	-684	-463	-388	-555	-466
I_{corr}	After 4 cycles	0.71	0.18	0.22	0.09	0.07

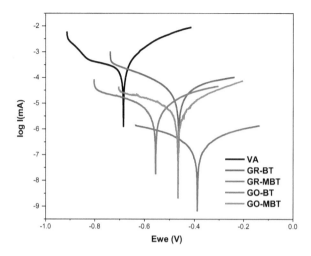

Figure 4.10 Tafel plots of the vinyl acrylate coating and carbon containers incorporated vinyl acrylate coatings after 4 cycles of UV weathering analysis.

acrylate coating (-684 mV). A maximum shift was noticed for the VA/Gr-MBT and VA/Gr-BT coatings to -388 mV and -463 mV, respectively. The VA/GO-BT and VA/GO-MBT coatings also showed a positive shift of -555 mV and -466 mV respectively as compared with E_{corr} of the pure VA coating. The I_{corr} values of the carbon containers-based coatings were noted to be lower than I_{corr} of the pure vinyl acrylate coating. These results indicated that both graphene and graphene oxide-based coatings provided an effective UV protection. Such coatings can be a beneficial alternative to the existing silica nanocontainer coatings which provide protection only by releasing corrosion inhibitor from the polyelectrolyte shell. On the other hand, in addition to the active protection by the corrosion inhibitor released from the carbon containers, the carbon materials can provide a passive protection as well. These materials retain the integrity of the coatings by resisting the UV degradation as well as preventing the unwanted release of the corrosion inhibitor into the coating. The carbonaceous materials also provide the protection by increasing the tortuous path of electrolyte diffusion through the coatings.

4.4 Conclusion

The carbon based containers were prepared by layer by layer addition of the polyelectrolytes and corrosion inhibitor onto the surface of graphene and graphene oxide. The containers were prepared based on graphene and graphene oxide as host matrices and encapsulating two different corrosion inhibitors (benzotriazole and 2-mercaptobenzotriazole) in the polyelectrolyte shells. The GO-BT and GO-MBT containers exhibited an agglomeration tendency during the synthesis of coatings with vinyl acrylate matrix which led to a slightly decreased performance of these coatings as compared with the Gr-BT and Gr-MBT containers-based coatings. Overall, the carbon container coatings were active in providing protection over 2 weeks during immersion in 0.35 M sodium chloride solution. The irregularities formed due to the incorporation of the carbon containers resulted an initial decrease in the coating resistance (R_c) and charge transfer resistance (R_{ct}). However, the corrosion inhibitor released from the carbon containers provided a self-healing nature to the coatings by forming a protective layer on the surface of metal that enhanced the R_c and R_{ct} values. At the same time, the self-healing nature of the carbon containers-based coatings was further supported by the anodic shift of E_{corr} and a lowering of I_{corr} of the coatings as compared with the pure VA coating. Further, the graphene-based containers provided a superior protection than the graphene oxide containers due to the uniform distribution of the graphene containers in the VA matrix. Furthermore, the UV protection of graphene and graphene oxide was not diminished by transforming them as the corrosion inhibitor carriers. It can be concluded from these studies that graphene and graphene oxide are beneficial alternatives to the silica matrices for preparing the corrosion inhibitor encapsulated containers for the active protection of the metal substrates.

Main Findings

The carbon-based containers encapsulated with corrosion inhibitor in the polyelectrolyte shell were fabricated using graphene (Gr) and graphene oxide (GO) as the host matrices. Four different containers were prepared encapsulating two different corrosion inhibitors, benzotriazole and 2-mercaptobenzotriazole, in the polyelectrolyte shells. An active protection of the coatings by the release of the corrosion inhibitor from the carbon containers was noticed over a period of two weeks in 0.35 M sodium chloride solution. The incorporation of carbon containers in the VA matrix diminished the electrochemical parameters (coating resistance (R_c) and charge transfer resistance (R_{ct})) at the beginning of the analysis. However, the tortuous path of diffusion of the electrolytes through the coatings and protective layer formed on the metal surface by the reaction of corrosion inhibitor released from the carbon containers with the metal ions provided a self-healing protection. It was attributed to an increase in the R_c and R_{ct} values during immersion in sodium chloride solution. The self-healing nature of the carbon containers-based coatings was further confirmed from an anodic shift of E_{corr} and a lowering of I_{corr} during the potentiodynamic measurements. The agglomeration of GO containers in the VA matrix diminished the protective behavior of the VA/GO-BT and VA/GO-MBT coatings as compared with the Gr based coatings. The incorporation of carbon containers imparted an effective UV protection to the VA coatings. Therefore, the organic coatings containing the corrosion inhibitor encapsulated carbon containers represent a functional methodology for providing an active and passive protection to the metal structures.

References

1. Nguyen, T. N., Hubbard, J. B., and McFadden, G. B. (1991) Mathematical model for the cathodic blistering of organic coatings on steel immersed in electrolytes. *Journal of Coatings Technology*, **63**, 43-52.
2. Zheludkevich, M. L., Serra, R., Montemor, M. F., Salvado, I. M. M., and Ferreira, M. G. S. (2006) Corrosion protective properties of nanostructured sol–gel hybrid coatings to AA2024-T3. *Surface Coating and Technology*, **200**, 3084-3094.
3. Ghosh, S. K. (2009) Self-healing polymers and polymer composites. In: *Self-healing Materials: Fundamentals, Design Strategies and Applications*, Ghosh, S. K. (ed.), Wiley-VCH, Germany, pp. 29-72.
4. Shi, X., and Dalal, N. S. (1994) Generation of hydroxyl radical by chromate in biologically relevant systems: role of Cr(V) complexes versus tetraperoxochromate(V). *Environmental Health Perspectives*, **102**, 231-236.
5. Ghosh, S. K. (2006) Functional coatings and microencapsulation: A general perspective. In: *Functional Coatings*, Ghosh, S. K., Wiley-VCH, Germany, pp. 1-28.
6. Abdullayev, E., Price, R., Shchukin, D., and Lvov, Y. (2009) Halloysite tubes as nanocontainers for anticorrosion coating with benzotriazole. *ACS Applied Materials & Interfaces*, **1**(7), 1437-1443.
7. Borisova, D., Akçakayıran, D., Schenderlein, M., Möhwald, H., and Shchukin, D. G. (2013) Nanocontainer-based anticorrosive coatings: Effect of the container size on the self-healing performance. *Advanced Functional Materials*, **23**, 3799-3812.

8. Shchukin, D. G., Zheludkevich, M., Yasakau, K., Lamaka, S., Ferreira, M. G. S., and Möhwald, H. (2006) Layer-by-layer assembled nanocontainers for self-healing corrosion protection. *Advanced Materials*, **18**, 1672-1678.
9. Zheludkevich, M. L., Tedim, J., and Ferreira, M. G. S. (2012) "Smart" coatings for active corrosion protection based on multi-functional micro and nanocontainers. *Electrochimica Acta*, **82**, 314-323.
10. Shchukin, D. G., and Moehwald, H. (2011) Smart nanocontainers as depot media for feedback active coatings. *Chemical Communications*, **47**, 8730-8739.
11. Ramezanzadeh, B., Ahmadi, A., and Mahdavian, M. (2016) Enhancement of the corrosion protection performance and cathodic delamination resistance of epoxy coating through treatment of steel substrate by a novel nanometric sol-gel based silane composite film filled with functionalized graphene oxide nanosheets. *Corrosion Science*, **109**, 182-205.
12. Fix, D., Andreeva, D. V., Lvov, Y. M., Shchukin, D. G., and Moehwald. H. (2009) Application of inhibitor-loaded halloysite nanotubes in active anti-corrosive coatings. *Advanced Functional Materials*, **19**, 1720-1727.
13. Abdullayev, E, and Lvov, Y. (2010) Clay nanotubes for corrosion inhibitor encapsulation: release control with end stoppers. *Journal of Materials Chemistry*, **20**, 6681-6687.
14. Tedim, J., Zheludkevich, M. L., Salak, A. N., Lisenkov, A., and Ferreira, M. G. S. (2011) Nanostructured LDH-container layer with active protection functionality. *Journal of Materials Chemistry*, **21**, 15464-15470.
15. Zheludkevich, M. L., Poznyak, S. K., Rodrigues, L. M., Raps, D., Hack, T., Dick, L. F., Nunes, T., and Ferreira, M. G. S. (2010) Active protection coatings with layered double hydroxide nanocontainers of corrosion inhibitor. *Corrosion Science*, **52**, 602-611.
16. Shchukin, D. G., Grigoriev, D. O., and Möhwald, H. (2010) Application of smart organic nanocontainers in feedback active coatings. *Soft Matter*, **6**, 720-725.
17. Chen, T., and Fu, J. J. (2012) An intelligent anticorrosion coating based on pH-responsive supramolecular nanocontainers. *Nanotechnology*, **23**, 505705.
18. Shchukin, D. G., Sukhorukov, G. B., and Möhwald, H. (2003) Smart inorganic/organic nanocomposite hollow microcapsules. *Angewandte Chemie, International Edition*, **42**, 4472-4475.
19. Zheludkevich, M. L., Shchukin, D. G., Yasakau, K. A., Möhwald, H., and Ferreira, M. G. S. (2007) Anticorrosion coatings with self-healing effect based on nanocontainers impregnated with corrosion inhibitor. *Chemistry of Materials*, **19**, 402-411.
20. Grigoriev, D. O., Kohler, K., Skorb, E., Shchukin, D. G., and Möhwald,

H. (2009) Polyelectrolyte complexes as a "smart" depot for self-healing anticorrosion coatings. *Soft Matter*, **5**, 1426-1432.
21. Skorb, E. V., Fix, D., Andreeva, D. V., Möhwald, H., and Shchukin, D. G. (2009) Surface-modified mesoporous SiO2 containers for corrosion protection. *Advanced Functional Materials*, **19**, 2373-2379.
22. Shchukin, D. G., and Mohwald, H. (2007) Surface-engineered nanocontainers for entrapment of corrosion inhibitor. *Advanced Functional Materials*, **17**, 1451-1458.
23. Dennis, R. V., Viyannalage, L. T., Gaikwad, A. V., Rout, T. K., and Banerjee, S. (2013) Graphene nanocomposite coatings for protecting low- alloy steels from corrosion. *American Ceramic Society Bulletin*, **92**, 18-24.
24. Li, Y., Yang, Z., Qiu, H., Dai, Y., Zheng, Q., Li, J., and Yang, J. (2014) Self-aligned graphene as anticorrosive barrier in waterborne polyurethane composite coatings. *Journal of Materials Chemistry A*, **2**, 14139-14145.
25. Chang, K.-C., Hsu, M.-H., Lu, H.-I., Lai, M.-C., Liu, P.-J., Hsu, C.-H., Ji, W.-F., Chuang, T.-L., Wei, Y., Yeh, J.-M., and Liu, W.-R. (2014) Room-temperature cured hydrophobic epoxy/graphene composites as corrosion inhibitor for cold-rolled steel. *Carbon*, **66**, 144-153.
26. Chang, C.-H., Huang, T.-C., Peng, C.-W., Yeh, T.-C., Lu, H.-I., Hung, W.-I., Weng, C.-J., Yang, T.-I., and Yeh, J.-M. (2012) Novel anticorrosion coatings prepared from polyaniline/graphene composites. *Carbon*, **50**, 5044-5051.
27. Chang, K., Hsu, C., Lu, H., Ji, W., Chang, C., Li, W., Chuang, T., Yeh, J., Liu, W., and Tsai, M. (2014) Advanced anticorrosive coatings prepared from electroactive polyimide/graphene nanocomposites with synergistic effects of redox catalytic capability and gas barrier properties. *Express Polymer Letters*, **8**, 243-255.
28. Liu, S., Gu, L., Zhao, H., Chen, J., and Yu, H. (2015) Corrosion resistance of graphene-reinforced waterborne epoxy coatings. *Journal of Materials Science and Technology*, **32**, 425-431.
29. Zhang, Z., Zhang, W., Li, D., Sun, Y., Wang, Z., Hou, C., Chen, L., Cao, Y., and Liu, Y. (2015) Mechanical and anticorrosive properties of graphene/epoxy resin composites coating prepared by in-situ method. *International Journal of Molecular Sciences*, **16**, 2239-2251.
30. Dumée, L. F., He, L., Wang, Z., Sheath, P., Xiong, J., Feng, C., Tan, M. Y., She, F., Duke, M., Gray, S., Pacheco, A., Hodgson, P., Majumder, M., and Kong, L. (2015) Growth of nano-textured graphene coatings across highly porous stainless steel supports towards corrosion resistant coatings. *Carbon*, **87**, 395-408.
31. Chaudhry, A., Mittal, V., and Mishra, B. (2015) Inhibition and promotion of electrochemical reactions by graphene in organic coatings. *RSC Advances*, **5**, 80365-80368.
32. Mayavan, S., Siva, T., and Sathiyanarayanan, S. (2013) Graphene ink

as a corrosion inhibiting blanket for iron in an aggressive chloride environment. *RSC Advances*, **3**, 24868-24871.
33. Chen, T., and Fu, J. J. (2012) An intelligent anticorrosion coating based on pH-responsive supramolecular nanocontainers. *Nanotechnology*, **23**, 505705.
34. Cui, J., Liu, Y., and Hao, J. (2009) Multiwalled carbon-nanotube-embedded microcapsules and their electrochemical behavior. *The Journal of Physical Chemistry C*, **113**(10), 3967-3972.
35. Vengatesan, M. R., Singh, S., Stephen, S., Prasanna, K., Lee, C. W., and Mittal, V. (2017) Facile synthesis of thermally reduced graphene oxide-sepiolite nanohybrid via intercalation and thermal reduction method. *Applied Clay Science*, **135**, 510-515.
36. Chou, T. P., Chandrasekaran, C., Limmer, S., Nguyen, C., and Cao, G. Z. (2002) Organic-inorganic sol-gel coating for corrosion protection of stainless steel. *Journal of Materials Science Letters*, **21**, 251-255.
37. Brinkerandg, C. J., and Scherer, W. (1990) *Sol-Gel Science: The Physics and Chemistry of Sol-Gel Processing*, Academic Press, USA.
38. Pierre, A. C. (1998) *Introduction to Sol-Gel Processing*, Kluwer, USA.
39. *Ultrastructure Processing of Glasses, Ceramics and Composites*, Hench, L. L., and Ulrich, D. R. (eds.), Wiley, USA (1984).
40. Verma, S., and Dutta, R. K. (2015) A facile method of synthesizing ammonia modified graphene oxide for efficient removal of uranyl ions from aqueous medium. *RSC Advances*, **5**, 77192-77203.
41. Pham, V. H., Cuong, T. V., Hur, S. H., Oh, E., Kim, E. J., Shin, E. W., and Chung, J. S. (2011) Chemical functionalization of graphene sheets by solvothermal reduction of a graphene oxide suspension in N-methyl-2-pyrrolidone. *Journal of Materials Chemistry*, **21**, 3371-3377.
42. Kraljic, M., Mandic, Z., and Duic, L. (2003) Inhibition of steel corrosion by polyaniline coatings. *Corrosion Science*, **45**, 181-198.
43. Pereira da Silva, J. E., Cordoba de Torresi, S. I., and Torresi, R. M. (2007) Polyaniline/poly(methylmethacrylate) blends for corrosion protection: The effect of passivating dopants on different metals. *Progress in Organic Coatings*, **58**, 33-39.
44. Akid, R., Gobara, M., and Wang, H. (2011) Corrosion protection performance of novel hybrid polyaniline/sol–gel coatings on an aluminium 2024 alloy in neutral, alkaline and acidic solutions. *Electrochimica Acta*, **56**, 2483-2492.
45. Conceicao, T. F., Scharnagl, N., Blawert, C., Dietzel, W., and Kainer, K. U. (2010) Corrosion protection of magnesium alloy AZ31 sheets by spin coating process with poly(ether imide). *Corrosion Science*, **52**, 2066-2079.
46. Latnikova, A., Grigoriev, D., Schenderlein, M., Mohwald, H., and Shchukin, D. (2012) A new approach towards "active" self-healing coatings: exploitation of microgels. *Soft Matter*, **8**, 10837-10844.
47. Sugama, T., and Cook, M. (2000) Poly(itaconic acid)-modified chito-

san coatings for mitigating corrosion of aluminum substrates. *Progress in Organic Coatings*, **38**, 79-87.
48. Zheludkevich, M. L., Serra, R., Montemor, M. F., Yasakau, K. A., Salvado, I. M. M., and Ferreira, M. G. S. (2005) Nanostructured sol–gel coatings doped with cerium nitrate as pre-treatments for AA2024-T3: Corrosion protection performance. *Electrochimica Acta*, **51**, 208-217.
49. Latnikova, A., and Yildirim, A. (2015) Thermally induced release from polymeric microparticles with liquid core: the mechanism. *Soft Matter*, **11**, 2008-2017.
50. Chongdar, S., Gunasekaran, G., and Kumar, P. (2005) Corrosion inhibition of mild steel by aerobic biofilm. *Electrochimica Acta*, **50**(24), 4655-4665.

5

Self-healing Protective Coatings of Polyvinyl Butyral/Polypyrrole-Carbon Black Composite on Carbon Steel

5.1 Introduction

Intrinsically conducting polymers (ICPs) have attracted much research interest in the area of protective coatings because of their properties such as self-healing and environmental friendliness, which make them promising candidates to replace the coatings based on hexavalent chromium [1-3]. Several possible corrosion inhibition mechanisms have been reported for ICPs, though there still exists some ambiguity concerning the exact protection pathway. The most cited mechanism is the surface ennobling and anodic passivation of steel induced by the inherent redox capability of conducting polymers [4-6]. Another proposed mechanism is the so-called self-healing, where ICPs doped with suitable anionic inhibitors act as a reservoir of corrosion inhibitors. On initiation of corrosion, oxidation of the substrate changes the oxidation states of these conducting polymers which forces the doping agent to be released at the affected area [7-13]. These smart coatings are able to prevent corrosion even in the scratched areas where bare steel comes in direct contact with the aggressive environment [7,10].

Among the different ICPs, polypyrrole (PPy) is promising in terms of its high conductivity, stability and ease of synthesis [14,15]. However, some inherent characteristics of PPy, such as its poor mechanical properties, porosity and poor solubility in common solvents, have limited its application in coatings [4]. Also, it is difficult to cast PPy films due to its poor solubility in solvents resulting from the strong intermolecular attraction between PPy chains. Direct deposition either by chemical or electrochemical methods was commonly used for the fabrication of PPy based protective coatings [8,9,11,12]. Recently, a new strategy for corrosion protection by using conducting polymers as additive components or inhibiting pigments in composite coatings has been developed, by which the undesirable properties of

Thanapoon Niratiwongkorn, Gisha Elizabeth Luckachan and Vikas Mittal, Khalifa University of Science and Technology, Abu Dhabi, UAE
© 2021 Central West Publishing, Australia

both ICPs and organic polymers can be minimized [4,7,10,13-15]. A number of literature studies have reported the use of polyaniline (PANI) with other additives as inhibitive agents in commercial coatings [16-22]. However, a limited number of studies have been reported describing the use of PPy as inhibitive pigment [4,7,23,24]. Yan et al. [4] used oxyanion doped PPy deposited on aluminum flakes as an inhibitive pigment in epoxy primer for aluminum. PPy functioned both as a reservoir of corrosion inhibitor and an oxygen scavenger, thus, resulting in the self-healing of the artificially created defects on the epoxy coatings.

Corrosion protection by organic coatings is brought about by the barrier effect of the coating as well as internal sacrificial electrode formation, which protects the underlying metal from further corrosion [17]. The barrier properties of the coating can also be further enhanced by using suitable reinforcing fillers [18,25]. The addition of filler is generally reported to improve the adhesion strength between the polymer and metal as well as to increase the diffusion paths of water and oxygen molecules through the coating, thus, enhancing the corrosion resistance of the coating [17,26]. Graphene, a two-dimensional sheet of sp^2 carbon atoms, was recently reported as an ideal filler for corrosion inhibition in reinforced polymer coatings because of its barrier nature towards corrosion promoting species [26-35]. Recently, Chaudhry et al. [27] discussed the anti-corrosion behavior of graphene incorporated into a self-cross linked polyvinyl butyral composite. Graphene platelets increased the diffusion path of oxygen and ionic species through the coating, which improved corrosion resistance of the coating. Similar mechanisms were reported for graphene/epoxy [28,29], polyurethane/graphene [30] and poly(sodium styrene sulfonate)/graphene [31] composites as protective coatings. The use of polymer composites consisting of graphene and conducting polymers presents a new approach for making corrosion resistant coatings. The electrical conductivity of graphene imparts corrosion protection by enhancing the passive oxide layer formation at metal polymer interface, in addition to its diffusion barrier nature. Sreevatsa et al. [32] investigated the potential of graphene as an ionic barrier for steel in conjunction with a polypyrrole 'topcoat' to form a p–n junction. Chang et al. [33] reported the application of polyaniline/graphene composites (PAGCs) for corrosion protection of steel.

In the present study, fabrication of stable self-healing organic coatings was reported by incorporating polypyrrole-carbon black (PPyCB) composite pigment into a polyvinyl butyral (PVB) matrix.

The effect on the corrosion protection of PVB/PPyCB composite coatings by the addition of conducting particles like graphene was also analyzed. The protective nature of these coatings was studied by immersion test and electro- chemical methods.

5.2 Experimental

5.2.1 Materials

Polypyrrole (doped, conductivity 30 S cm^{-1} (bulk), extent of labeling: 20 wt% loading, composite with carbon black, melting point > 300 °C), polyvinyl butyral with trade name Butvar B-98 (molecular weight 40,000–70,000 and specific gravity of 1.1 at 23 °C), methanol and hydrochloric acid were purchased from Sigma-Aldrich. Graphene nanoplatelets (with a trade name of N002-PDR) were purchased from Angstron materials, USA, and were used as received.

5.2.2 Preparation of PVB/PPyCB Formulations

The required amount of PVB was dissolved in 20 mL methanol by stirring for 6 h. To avoid the evaporation of methanol, coating formulations were prepared in round bottom flask fixed with a water condenser. The methanol level on the flask was marked after the addition of PVB and was maintained during the stirring and sonication processes. After dissolving PVB completely in methanol, the PPyCB composite pigment was added slowly with magnetic stirring. The total amount of PVB and PPyCB composite was fixed at 2 g. The mixture was subjected to continuous stirring for 12 h, followed by 8 h sonication to obtain a well dispersed PVB/PPyCB composite formulation. Three different formulations were prepared by using 5%, 10% and 20% PPyCB composite in the PVB matrix (denoted as PVB/ PPyCB5, PVB/PPyCB10 and PVB/PPyCB20). Table 5.1 shows the amounts of different components in the PVB/PPyCB composite formulations. In order to incorporate graphene in the PVB/PPyCB composite, 5 wt% of graphene was added into the PVB/PPyCB20 formulation and sonicated for 12 h followed by stirring for another 6 h.

5.2.3 Application of Composite Formulations on Carbon Steel

Carbon steel of grade RST37-2 DIN 17100-80 was purchased from the Qatar Steel Industries Factory, Qatar. Its composition in weight%

was C (0.125), Mn (0.519), Si (0.016), P (0.014), S (0.005), Al (0.034) and Fe (99.287). Carbon steel coupons of 5 cm x 2 cm x 2 cm were first pickled with hydrochloric acid for 2 h in order to remove the oxide layer from the surface. After acid treatment, the coupons were polished with sandpaper (60, 150 and 180 grits) followed by water and acetone rinsing. Finally, the coupons were sonicated in acetone for 10 min and dried in oven at 90 °C for 1 h. The prepared PVB/PPyCB formulations were applied on the metal coupons by dip coating using QPI-128 dip coater from Qualtech Product Industry Co., Ltd. The metal coupons were dipped in the solution with an immersion rate of 100 mm s^{-1} for 30 s, and then withdrawn from the solution with the same rate as immersion. This process was repeated 3 times to obtain a suitable coating thickness. After the coating, the metal coupons were cured in the oven at 190 °C for 1 h to obtain cross-linking between PVB, PPyCB and metal surface, so as to improve adhesion. After curing, the metal coupons were cooled down to room temperature. The thickness of each coating was measured using PosiTector 6000 coating thickness gages from DeFelsko Corporation, New York. Adhesion strength of the coatings was measured by the Cross Cut Tape Test, according to the ASTM standard test method D3359-09.

Table 5.1 Amount of different components used in the PVB/PPyCB composite formulations

Sample	PVB (g)	PPyCB (g)	Total (g)
PVB	2.00	-	2.00
PVB/PPyCB5	1.90	0.10	2.00
PVB/PPyCB10	1.80	0.20	2.00
PVB/PPyCB20	1.60	0.40	2.00

5.2.4 Electrochemical Measurements

Open circuit potential (OCP), electrochemical impedance spectroscopy (EIS) and polarization measurements were carried out at room temperature in a three-electrode corrosion cell consisting of a saturated calomel reference electrode (SCE), a platinum counter electrode and composite coated steel coupons as the working electrode. The electrolyte solution used for the electrochemical analysis was 4 wt% NaCl solution. The exposed area of working electrode was 1 cm^2. All measurements were performed on an electrochemical analyzer (supplied by BioLogic, France) using a software EC-Lab V10.39. Ultra-low current cables connected to the potentiostat were used for the accurate measurement of the current. This option includes current ranges from 100 nA down to 100 pA with additional gains extending the current ranges to 10 pA and 1 pA. The resolution on the lowest range was 76 aA. Open circuit potential was measured continuously for 48 h without disturbing the experimental set up, and impedance measurements were incorporated in-between the OCP cycles at regular intervals. Impedance measurements were performed as a function of open circuit potential by applying a sinusoidal potential in frequencies from 10^4 Hz to 10^{-2} Hz, with an amplitude of 15 mV. The same software was used to simulate the impedance behavior of the samples. Polarization measurements were performed by polarizing the working electrode from an initial potential of -250 mV up to a final potential of +250 mV as a function of open circuit potential. The values of the corrosion current density (I_{corr}), corrosion potential (E_{corr}), and cathodic (β_c) Tafel constant were extracted from the polarization curves by Tafel extrapolation of the cathodic branch [36-38]. A scan rate of 0.1667 mV s^{-1} was used for the polarization sweep. To study the reproducibility of the measurement, each set of experiments was repeated three times on newly coated samples.

5.2.5 Immersion Test

The edges of the coated coupons were sealed using a Nippon epoxy primer. After 24 h drying at room temperature, the surface of the samples was scratched in an X shape using a surgical blade fixed into a cutter (with an ergonomic handle/holder) included with an adhesive test kit from GARDCO Paul N. Gardner Co., Inc. USA. This special design of holder gave comfortable and precision control and helped to make identical scratches on all the samples. Scratched coupons immersed in 4 wt% NaCl solution were followed up daily using an optical microscope to monitor the self-healing effect of the coatings.

5.2.6 Characterization Techniques

The microstructure of the passive layer formed on the scratches was analyzed using a scanning electron microscope (SEM) (FEI Quanta, FEG250, USA) at an accelerating voltage of 5 kV. The chemical structure of the passive layer formed on the scratched areas was also analyzed using Raman spectroscopy. Raman spectra were recorded using a LabRAM HR spectrometer (Horiba Jobin Yvon). Laser light from a He/Ne source with a wavelength of 633 nm was used for excitation. A long working distance objective with magnification 50x was used to collect the scattered light as well as to focus the laser beam on the sample surface. A digital microscope, KH-7700 (Hirox Co. Ltd, USA), was used for the optical images of the coatings. Tensile testing was carried out using an Instron 2519-107 universal testing machine (Instron Corp.), according to the ASTM standard test method D6382V. Testing was done at room temperature using plain rubber grips, and the average of three replicas is reported here.

5.3 Results and Discussion

5.3.1 Morphology of the PVB/PPyCB Composite Coatings

To generate stable organic coatings of PVB/PPyCB composites on the carbon steel substrate, curing of coatings at different temperatures was carried out at the conventionally used ambient conditions [25] for generating the PVB coatings that did not provide corrosion protection on the incorporation of PPyCB pigments. These coatings were non-uniform and started to peel out while immersed in corrosive media. Curing for 1 h at 190 °C was observed to be optimal for attaining stable coatings of 20±5 μm thickness. The morphology of the composite coatings was compared through optical microscopy in Figure 5.1. The coatings exhibited uniform dispersion of the PPyCB pigment in the PVB matrix. The PVB/PPyCB5 coatings were smooth with no cracks on the surface (Figure 5.1a). However, as the amount of PPyCB was increased, the coatings became more prone to the appearance of surface cracks (Figures 5.1b and 5.1c). This might have occurred due to the increased amount of carbon black particles in the composite

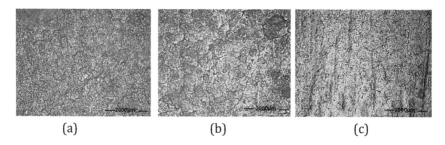

(a)   (b)   (c)

Figure 5.1 Texture of PVB/PPyCB composite coatings (a) PVB/PPyCB5, (b) PVB/PPyCB10, and (c) PVB/PPyCB20.

coatings which hindered a continuous film formation by reducing the contact between PVB and PPy.

The mechanical properties of PVB/PPyCB composites studied in this work revealed that all the composite films are brittle materials with high Young's moduli (E) and low elongations at break ($\varepsilon_{\sigma max}$). The Young's modulus of the PVB films modified with PPyCB increased with the concentration of conducting pigments. The elongation at break and tensile strength decreased in the PVB/PPyCB composites, though these mechanical parameters were independent of the PPyCB concentration which showed an irregular behavior at high concentrations (Table 5.2). However, a strong adhesion of PVB and PPyCB pigments on the metal surface occurred during the fabrication of the

Table 5.2 Tensile properties of the pure PVB and PVB/PPyCB composites

Sample	Young's modulus (E) MPa	Tensile strength (σ_{max}) MPa	Elongation at break ($\varepsilon_{\sigma max}$) %
PVB	1181±190	32.0±8	2.77±0.1
PVB/PPyCB5	1838±220	29.9±10	2.23±0.4
PVB/PPyCB10	2134±188	23.5±9	1.65±0.2
PVB/PPyCB20	2170±325	25.1±7	1.70±0.2

PVB/PPyCB coatings at high temperature. Figure 5.2 shows the images of crosscut tape test conducted on PVB/ PPyCB20 coatings before and after 1 d immersion in 4 wt% sodium chloride solution. The

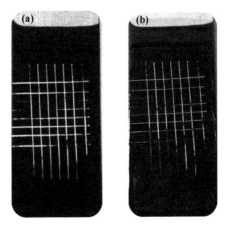

Figure 5.2 Crosscut tape test on PVB/PPyCB20 composite coating (a) before and (b) after 2 d immersion in 4 wt% NaCl solution.

edges of the cut in Figure 5.2a were smooth and the squares of the lattices were not detached, indicating the strong adhesion of the composite matrix on the metal surface. During the immersion, even though the corrosion products were formed mainly at the crosscut regions, most of the coating areas were not detached while the tape test (Figure 5.2b), which showed the stability of the PVB/PPyCB20 coating, indicated that strong binding on the metal surface prevented the ingress of water through the edges of the scratches.

5.3.2 Electrochemical Measurements

The corrosion resistance of the PVB/PPyCB composite coatings was studied by electrochemical measurements as both metals and conducting polymers lead to redox reactions when in contact with electrolyte. OCP and polarization measurements were conducted on the PVB and PVB/PPyCB composite coatings by exposing a 1 cm² area to 4 wt% NaCl solution. Evolution of OCP with time of immersion is shown in Figure 5.3. The initial OCP values of all coatings were observed to be more positive than that of bare steel (-0.6 V vs. SCE) [8], thus confirming the barrier nature of the coatings. The OCP of the PVB

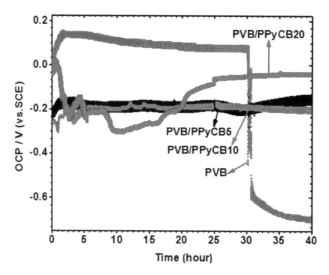

Figure 5.3 OCP vs. time plots of PVB and PVB/PPyCB composite coatings in 4 wt% NaCl solution.

coating exhibited a passive potential of steel at 0.05 V vs. SCE. After 30 h of immersion, it decreased sharply to a potential (-0.66 V vs. SCE) close to the corrosion potential of bare steel and remained in the same range till the completion of measurement. Thus, protection of the metal surface by the PVB coating persisted over a short time period and as corrosion started on metal surface by the ingress of electrolyte through the defects in the coating, OCP decreased rapidly to the active dissolution potential of iron. Since PVB is not an inhibitive polymer, the corrosion initiated on the metal surface continued to propagate and no further corrosion protection was offered by the PVB coating. After a slight reduction in the OCP value from -0.12 V vs. SCE to -0.22 V vs. SCE, 5% the PPyCB coating exhibited a shift in the potential to -0.15 V vs. SCE after 1 h immersion, which was retained till the end of the 2 d analysis. A similar trend was observed for the 10% PPyCB coating. After the initial decrease from -0.26 V vs. SCE to -0.30 V vs. SCE, OCP required 2 h to reach plateau behavior at -0.19 V vs. SCE. Unlike the PVB coating, the tendency of PPyCB coatings to shift the OCP to potentials nobler than bare steel increased with enhancing PPyCB content in the composite coating. The initial passive potential of the PVB/PPyCB20 composite coating at -0.03 V vs. SCE decreased to a pseudo-plateau at -0.20 V vs. SCE and was retained for 10 h. At this plateau stage, some oscillations of the potential could be observed, which were attributed to the anion exchange of the PPy coating with the environment during equilibration in NaCl solution [12]. Such oscillations were also observed for the 5% and 10% PPyCB coatings, however, the time taken for the stabilization was lower than for the 20% PPyCB coating. After the stabilization, the potential for 20% PPyCB coating was decreased to the dissolution potential of iron at -0.30 V vs. SCE. Two hours after the start of corrosion, OCP increased slowly to a plateau value of -0.04 V vs. SCE, which was close to the initial potential, thus indicating that metal subjected to corrosion was shifted to a more noble side. The positive shift of OCP basically shows the passive state of underlying metal because of the good corrosion protection ability of the surface film [39]. Therefore, the potential jump of 260 mV was associated with re-passivation of the defects on the coatings which might have resulted due to the deposition of protective layers on the metal surface. Kowalski et al. [10] have also reported a similar behavior of self-healing on bi-layered PPy coatings doped with hetero-polyanions, which resulted due to the synergistic effect of PPy redox reactions and catalytic reaction of dopant anions with metal ions. The first potential plateau value at around -0.20 V vs. SCE of PVB/PPyCB20 coatings was associated with the extended passive state of the iron, which was noted by Hien et al.

[13] and Nguyen et al. [9] for analysis of PPy coated iron substrates. According to these reports, this potential also results in the decreased ingress of chloride anion in the coating. However, the first plateau, extended for a short duration, moved to a second potential plateau at -0.30 V vs. SCE. At this potential, active dissolution of iron with Cl⁻ attack and subsequent pit formation took place, which drove the polypyrrole into an un-doped state. The released dopant formed stable iron complexes inside the defect, and, thus, provided a blocking effect for further iron dissolution. The return of OCP towards the initial potential (-0.04 V vs. SCE) suggested that the deposited iron complex and the PPy polymer effectively blocked the diffusion of corrosive ions to the passive oxide layer formed beneath the deposit. It should be noted here that, before the recovery of noble potential, the steel remained in the active potential region in which iron dissolved and the duration of potential shift to the noble region was also high. Such a shift of OCP towards the noble potential of steel was also reported by Kowalski et al. [10], where a passive oxide layer was stabilized by the iron molybdate deposition at the defects.

Corrosion protection of PPyCB pigments in the composite coatings was studied further by measuring the impedance at the open circuit potential. 5% and 10% PPyCB composites showed low frequency impedance in the same order of PVB at 10^9 with a slightly decreased magnitude. However, Bode plots of the 20% PPyCB coating measured after 5 min immersion showed a higher low frequency impedance than the pure PVB coating (Figure 5.4a). Corrosion protection of organic coatings is mainly associated with their barrier properties, which prevent oxygen and moisture from reaching the metal surface [17]. Such barrier coatings generally show a time constant at high frequency regions in the EIS plots [40]. Therefore, the plateau in the high frequency region in the Bode phase angle plot of the PVB coating could be attributed to its barrier effect, which remained close to 90° over half of the measurement frequencies (10^1 to 10^4, Figure 5.4b). This plateau nature of the Bode phase angle plot was not affected by the addition of PPyCB pigment into the PVB matrix, even at a high concentration of 20% PPyCB composite. It suggests that the PVB/PPyCB composite coatings had a barrier nature, which was sufficient to protect the underlying metal from corrosion. Such a barrier nature would have resulted from good dispersion and better interaction of PPyCB pigments with the metal and PVB matrix. The low frequency phase angle at 45° of PVB/PPyCB20 composite coating in comparison to nearly zero (6.5°) of pure PVB coating shows the enhanced corrosion protection of the coatings at high content of PPyCB pigments (Figure 5.4b). Furthermore, 2 d immersion in 4 wt% NaCl solution decreased the corrosion resistance of pure PVB coating.

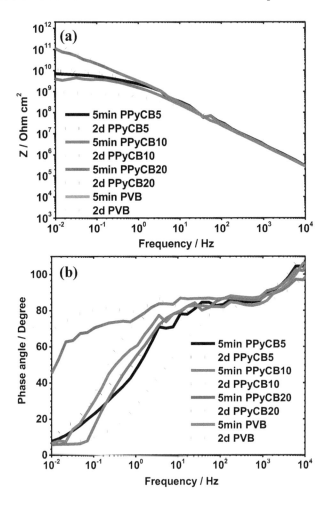

Figure 5.4 EIS plots of PVB and PVB/PPyCB composite coatings in 4 wt% NaCl solution: (a) Bode plot and (b) Bode phase angle plot.

It was observed in the Bode plot by the decrease in Z modulus at the low frequency region. Changes were more obvious in the Nyquist plot, where the diameter of the semicircle decreased significantly after 2 d immersion (Figure 5.5a). However, in the case of PPyCB composite coatings, low frequency Z modulus in the Bode plot increased after 2 d immersion (Figure 5.4a). Impedance in the Nyquist plot also increased significantly, especially at high concentrations of PPyCB composites (Figures 5.5a and 5.5b). Significant changes in the low frequency phase angle were also obvious in the Bode phase angle plot of PVB/PPyCB20 coating which shifted to 69° after 2 d exposure in sodium chloride solution (Figure 5.4b). These changes in the EIS plots indicated that PVB/PPyCB composite coatings imparted protection that occurred through the synergistic effect of redox reactions and electrical conductivity of the PPyCB pigments.

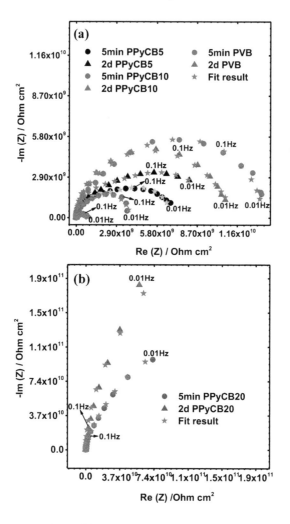

Figure 5.5 Nyquist plots of (a) PVB, PVB/PPyCB5, PVB/PPyCB10 composite coatings and (b) PVB/PPyCB20 composite coating after 5 min and 2 d immersion in 4 wt% NaCl solution.

Self-healing Protective Coatings of PVB/PPyCB Composite 149

To study the behavior of the PPyCB coatings in a corrosive environment, Nyquist plots were fitted with an equivalent circuit shown in Figure 5.6 and for which the obtained best fitting parameters are given in Table 5.3. The symbol R_s represents the solution resistance

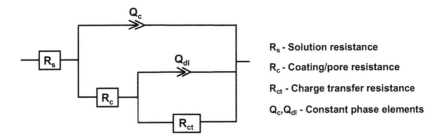

Figure 5.6 Equivalent circuit used for EIS modelling.

of the bulk electrolyte between the reference electrode and working electrode, R_c is the coating resistance, R_{ct} is the charge transfer resistance, Q_c and Q_{dl} are the constant phase elements (CPE) used instead of the pure capacitances C_c and C_{dl} respectively [41]. PVB exhibited R_c values typical for a barrier coating which decreased with immersion time. On the other hand, the continuous immersion increased the R_c values of the PPyCB coatings. Coating resistance (R_c) is related to the resistance of the electrolyte in pores, cracks and pits, and hence indicative of the barrier properties of the coating [42,43]. Therefore, the observed increase in R_c after 2 d immersion suggested that direct ingress of electrolyte through the coating is diminished. It can be further explained based on the constant phase element (CPE). According to the previous references, constant phase element permits the simulation of phenomenon that deviates from a pure capacitive behavior [19,20,41]. Therefore, the changes in CPE can be explained by evaluating the expression in terms of capacitance:

$C = Y_0(\omega_{max})^{n-1}$

where, Y_0 is the magnitude of the CPE, ω_{max} is the frequency at which the imaginary impedance reaches a maximum for the respective time constant, and n is the exponential term of the CPE which can vary between 1 for a pure capacitor and 0 for a pure resistor [19,20]. If n is equal to 1, the CPE behaves as a pure capacitor; when n is equal to 0, it represents a resistor and when n value is equal to -1, it represents

Table 5.3 EIS parameters of PVB and PVB/PPyCB composite coatings at different time of immersion in 4 wt% NaCl solution

	10 minutes							2 days						
			Q_c			Q_{dt}				Q_c			Q_{dt}	
Coatings	R_s^a (Ω cm²)	R_c^b (Ω cm²)	n^c	Y_o^d (Ω⁻¹ cm²)	R_{ct}^e (Ω cm²)	n^f	Y_o^g (Ω⁻¹ cm²)	R_s^h (Ω cm²)	R_r^i (Ω cm²)	n^j	Y_o^k (Ω⁻¹ cm²)	R_{ct}^l (Ω cm²)	n^m	Y_o^n (Ω⁻¹ cm²)
PVB	1447	9.27 × 10⁹	0.991	59.45 × 10⁻¹²	17.45 × 10⁹	0.989	52.80 × 10⁻¹²	1440	7.76 × 10⁵	0.977	0.116 × 10⁻⁹	1.20 × 10⁹	0.505	0.554 × 10⁻⁹
PVB/PPyCB5	705	1.73 × 10⁹	0.985	0.109 × 10⁻⁹	7.045 × 10⁹	0.802	0.156 × 10⁻⁹	2368	2.96 × 10⁹	0.999	0.080 × 10⁻⁹	8.60 × 10⁹	0.832	0.110 × 10⁻⁹
PVB/PPyCB10	1716	1.31 × 10⁹	0.978	0.101 × 10⁻⁹	3.154 × 10⁹	0.799	0.211 × 10⁻⁹	3775	8.67 × 10⁹	0.980	0.056 × 10⁻⁹	18.9 × 10⁹	0.825	0.168 × 10⁻⁹
PVB/PPyCB20	1016	0.19 × 10¹²	0.977	97.04 × 10⁻¹²	23.2 × 10⁹	0.818	4.37 × 10⁻⁹	3686	0.46 × 10¹²	0.999	82.25 × 10⁻¹²	0.22 × 10¹²	0.981	1.71 × 10⁻⁹

^{a} 5% probable error. ^{b} 9% probable error. ^{c} 1% probable error. ^{d} 8% probable error. ^{e} 1% probable error. ^{f} 12% probable error. ^{g} 2% probable error. ^{h} 3% probable error. ^{i} 7,8% probable error. ^{j} 3% probable error. ^{k} 10% probable error. ^{l} 1% probable error. ^{m} 10% probable error. ^{n} 11% probable error. ^{o} 1.5% probable error. ^{p} 5,6% probable error.

Self-healing Protective Coatings of PVB/PPyCB Composite

an inductor. Since the constant n given in the Table 5.2 is close to 1, CPE would be very similar to a pure capacitor, and hence, the constant Y_0 follows the same trend as the capacitance [44]. The capacitance of the coating (C_c) is proportional to its dielectric constant and can be, therefore, attributed to the amount of water absorbed by the coating [43]. The Q_c constant Y_0 of the PVB coating increased significantly after 2 d immersion in 4 wt% NaCl solution indicating the continuous ingress of electrolyte into the coating (Table 5.3). At the same time, PPyCB coatings showed a decrease in Y_0 values. Cascales et al. [45] also analyzed the polypyrrole/water interface using a molecular dynamics simulation study and stated that reduced polypyrrole was hydrophobic and oxidized polypyrrole was hydrophilic in nature. Due to the high hydrophobicity of the reduced polypyrrole, water molecules were repelled from the core of the polymer matrix and so did not penetrate into the polymer matrix. In the case of PPyCB composite coatings, the reduction of PPy by the diffusion of counter ions (un-doping) increased the hydrophobicity of the coating. As a consequence of this hydrophobicity, the water permeability of the coating decreased, and, thus, the total impedance of the system as well as its barrier nature increased. It was observed in the OCP plots in Figure 5.3 as retaining the barrier nature after the healing of defects on PVB/PPyCB20 coatings. In the PVB/PPyCB composite coatings, anodic dissolution of iron by the attack of corrosive species initiated undoping of polypyrrole, resulting in the formation of passive film at the defected areas which prevented corrosion. The diminished corrosion process of the metal surface by the passive layer deposition are obvious in the R_{ct} values. R_{ct}, corresponding to the resistance to charge transfer processes on the metal surface, increased significantly for 20% PPyCB coatings with immersion time, whereas it decreased for pure PVB coating (Table 5.3). During the process of the un-doping of the PPy, the oxidation charge of PPy changed to the reduced form. Ions from the electrolyte (Cl^- and OH^-) balanced the charge within the polymer [4,9,46,47]. At prolonged exposure time, the reduction of the polymer became the main cathodic reaction and the percentage of reduced polymer increased the hydrophobicity of the coating and, thus, decreased the water permeability [14]. The reduced PPy had an ability to capture dissolved oxygen in the coating, which was reported by Yan et al. [4]. This oxygen-scavenger effect of PPy pigments decreased the O_2 permeability of the coating and, subsequently the corrosion rate of the substrate. A schematic representation of corrosion protection provided by PVB/PPyCB composite coatings is shown in

Figure 5.7. As per the scheme, the redox properties of PPy provided an anodic protection to the steel by inducing a passive film formation at the defect area, which, in turn, increased the amount of reduced PPy near the passivated areas. This passive film along with the hydrophobic nature of the reduced PPy limited the charge transfer reaction rate by acting as a diffusion barrier.

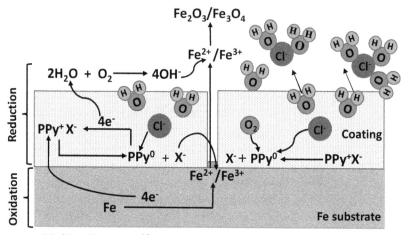

Figure 5.7 Schematic representation of corrosion protection imparted by PPy pigments in PVB/PPyCB composite coatings.

Figure 5.8 shows the cathodic and anodic polarization curves recorded for the PVB and PVB/PPyCB composite coatings after 1 d of exposure in 4 wt% NaCl solution. It is obvious in the plots that anodic polarization curves deviated from the linear Tafel behavior. This deviation would have occurred by the formation of passive film and pitting. The existence of passivation in conjunction with a dissolution reaction, due to pitting, does not result in a well-defined experimental anodic Tafel region [38,45]. Due to the absence of linearity in the anodic branch, accurate evaluation of the anodic Tafel slope by Tafel extrapolation of the anodic branch was not possible. Therefore, Tafel extrapolation of the cathodic branch of the polarization curve to the corrosion potential (E_{corr}) was used for the determination of I_{corr} [36]. Thus, the obtained electrochemical parameters, corrosion potential (E_{corr}), corrosion currents densities (I_{corr}) and cathodic Tafel slope (β_c) are tabulated in Table 5.4 as a function of PPyCB concentration in the composite coating. 5% and 10% PPyCB coatings showed E_{corr} in a more negative side than pure PVB coating after 1 d of immersion.

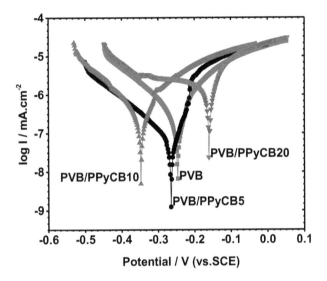

Figure 5.8 Polarization curves of PVB and PVB/PPyCB composite coatings after 1 d immersion in 4 wt% NaCl solution.

However, an anodic shift of E_{corr} occurred for 20% PPyCB coating, thus confirming the protection provided by PPyCB at high concentrations. The penetration of electrolyte down to the metal surface through the pores or cracks present on the coating initiated the corrosion processes. With time, PPy in the coating released dopant anions by redox reactions and suppressed the corrosion by passivating the defected areas on the metal surface, which shifted the corrosion

Table 5.4 List of electrochemical parameters obtained from polarization curves of PVB and PVB/PPyCB composite coatings after 1 d immersion in 4 wt% NaCl solution

Coatings	E_{corr} (V vs. SCE)	I_{corr} (mA cm^{-2})	β_c (mV per decade)
PVB	−0.246	3.60 × 10^{-7}	135
PVB/PPyCB5	−0.265	0.67 × 10^{-7}	137
PVB/PPyCB10	−0.348	3.88 × 10^{-7}	132
PVB/PPyCB20	−0.160	2.00 × 10^{-6}	747

potential of the metal to anodic side. The suppression of the anodic dissolution process by the deposited passive layers was attributed to the deviation of anodic polarization curves from the linear Tafel behavior. The deposited passive layer along with the reduced PPy hindered the diffusion of oxygen and water down to the metal surface which diminished the cathodic reduction reactions, as observed from the high value to cathodic Tafel slope (β_c) of 20% PPyCB after 1 d exposure (Table 5.4). Therefore, it can be suggested that PPy imparted anodic and cathodic protection to the underlying metal substrate. However, the corrosion current densities (I_{corr}) of the 20% PPyCB coating were higher than those of the pure PVB coating. The enhancement in I_{corr} might be caused by the contribution from PPy redox reactions in the PVB matrix, and it was observed to be increasing with increasing PPy content in the coatings. A similar increase in the corrosion current density of an aluminum substrate with a PPy coating has also been reported by Liu et al. [11]. Though the E_{corr} of 5% and 10% PPyCB coatings were more negative than the PVB coating, the Icorr of the 5% PPyCB coating was lower than PVB and nearly the same value as the PVB was observed for the 10% PPyCB coating, indicting the protection provided by these coatings as well. The increased I_{corr} of the 10% PPyCB coating in comparison to the 5% PPyCB coating would have resulted from the redox reactions of PPy. Though 5% and 10% PPyCB coatings also provided protection, this effect was not sustainable for longer periods of time due to the lesser amount of PPy in the PPyCB composite coatings. For comparison, the corrosion resistance behavior of other nano- composite coatings of PPy and polyaniline (PANI) reported previously are given in Table 5.5 [17,48-50]. The E_{corr} of these coatings was recorded more in the cathodic side and I_{corr} was higher than those noted for the PVB/PPyCB

Table 5.5 Comparison of E_{corr} and I_{corr} of other nanocomposite coatings of polypyrrole and polyaniline reported in references 17, 48-50

Material*	Coating thickness	Corrosive media	E_{corr} (V)	I_{corr} (mA.cm^{-2})
PVB/PANI	117	3.5 wt% NaCl	-0.30	-
MMT/PEA	20	5.0 wt% NaCl	-0.48	2.69x10-4
Na-MMT/PPy	-	3.5 wt% NaCl	-0.43	0.2827
Epoxy/PPy-flyash	-	3.5 wt% NaCl	-0.50	1.40x10-4

PVB: polyvinyl butyral, PANI: polyaniline, MMT: montmorillonite clay, PEA: poly(o-ethoxy aniline), PPy: polypyrrole."

composite coatings in Table 5.3. It clearly shows the enhanced resistance of PVB coating to corrosion processes of the underlying metal, because of the inhibition behavior of incorporated PPyCB pigments.

5.3.3 Immersion Test

In order to further analyze the protection of the PVB/PPyCB coatings and the nature of protective film formed beneath the coatings, defects were created intentionally on the coating surfaces, followed by immersion in 4 wt% NaCl solution. The images of the PPyCB composite coatings in Figure 5.9 were taken when the scratches became invisible. Optical images of the scratches on the coatings' surface indicated that the corrosion was stopped in composite coatings, specifically at high PPyCB pigment concentrations. Healing of the scratches on PPyCB coatings was identified by looking for the black and red colored products expected to form at the corroded areas, similar to the optical image of bare steel in Figure 5.9b. Optical images of the scratches on the PVB coating indicated complete corrosion with black corrosion products during the immersion in a 4 wt% NaCl solution. The PVB/PPyCB5 coating also exhibited a similar result as pure PVB coating, except for certain areas of the scratch which were covered by a passive film, instead of red and black corrosion products. The effect of passive film formation was more prominent in the PVB/PPyCB10 and PVB/PPyCB20 composite coatings. In addition, the time taken for the passive film formation was related to the PPyCB content in the coatings. For the PVB/PPyCB10 coating, passive film formation took about 3 d to cover most of the scratch, whereas the scratch was protected within 2 d for the PVB/PPyCB20 coating. Significant healing occurred for the PVB/PPyCB20 coating where the self-healed scratches were difficult to identify from the rest of the areas (Figure 5.9j). In addition, no red and black corrosion products were observed on the healed scratches of PVB/ PPyCB10 and PVB/PPyCB20 coatings (Figures 5.9h and 5.9j), which were observed on the scratches of PVB and PVB/PPyCB5 coatings (Figures 5.9d and 5.9f), indicating the enhanced protection of PVB/PPyCB composite coating at high percentages of PPy pigment. In order to analyze the nature of passive film, scratched areas were subjected to SEM and Raman spectroscopic measurements. Figures 5.10a and 5.10b show SEM images of scratches on the PVB surface and PVB/PPyCB20 composite coating respectively after 2 d immersion in 4 wt% NaCl solution. Corrosion

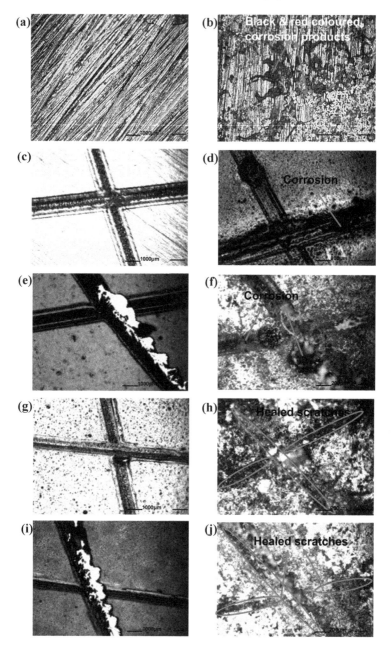

Figure 5.9 Optical images of scratches during immersion in 4 wt% NaCl solution. Bare steel (a) before and (b) after 2 d; PVB (c) before and (d) after 7 d; PVB/PPyCB5 (e) before and (f) after 4 d; PVB/PPyCB10 (g) before and (h) after 3 d; PVB/PPyCB20 (i) before and (j) after 3 d.

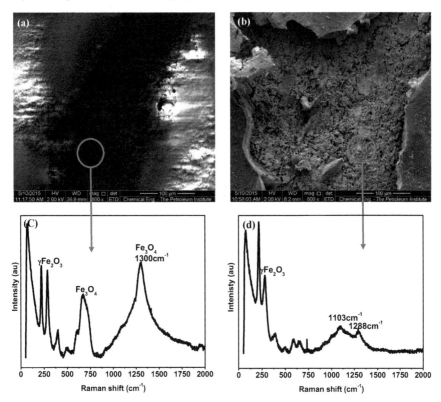

Figure 5.10 SEM images of scratches on (a) PVB coating and (b) PVB/PPyCB20 composite coating after 2 d immersion in 4 wt% NaCl solution. Raman spectra of marked areas on the scratches for (c) PVB coating and (d) PVB/PPyCB20 composite coating.

products distributed on the metal surface as agglomerated particles were obvious in the scratch on the PVB coating (Figure 5.10a). Raman spectra of these particles exhibited characteristic bands of Fe_3O_4 oxides at 405 cm^{-1}, 655 cm^{-1} and 1308 cm^{-1} and characteristic Raman shifts of γ-Fe_2O_3 at 223 cm^{-1} and 291 cm^{-1} (Figure 5.10c) [5,51]. The scratch on the PVB/PPyCB20 composite coating was observed to be completely covered by a thick layered material (Figure 5.10b). The Raman spectrum of this layer exhibited sharp signals in the region of γ-Fe_2O_3 oxides and less intense signals in the Fe_3O_4 oxides region (Figure 5.10d). In addition to these iron oxide signals, two broad Raman shifts were apparent at 1288 cm^{-1} and 1103 cm^{-1}, attributed to the asymmetric and symmetric axial deformation of O=S=O groups of sulfonic acid [11]. Thus, it can be assumed that the anodic dissolution of iron activated organic sulfonic acid doped PPy to release the dopant at the oxidizing areas, which created a stable complex with metal irons and stabilized the passive oxide layers on the metal surface, thus, preventing the underlying metal from further corrosion. Thus, from electrochemical and microscopic analysis, it should be concluded that doped PPy in the PVB/PPyCB composite coatings provided protection by acting as an inhibitive additive. In an earlier study, Armelin et al. [15] generated PPyCB composite coatings of approx. 140 µm thickness with epoxy matrix on carbon steel and PPy concentration was optimized to 1 wt% for the maximum corrosion protection. Ruhi et al. [23] also fabricated a three-component system of chitosan-polypyrrole-SiO_2 composite coatings with epoxy polymer and observed an adequate corrosion protection at a maximum of 2 wt% of chitosan-polypyrrole-SiO_2 pigment in epoxy matrix. However, in the case of PVB/PPyCB composite coatings, a maximum of 20 wt% was incorporated in the PVB matrix at a much lower coating thickness of 20±5 µm and a significant protection was observed at 48 h of immersion in 4 wt% NaCl solution.

5.3.4 Protective Performance of Graphene Incorporated PVB/PPyCB Composite Coatings

Figure 5.11a shows the changes in OCP of PVB/PPyCB20/Gr5 coating as a function of immersion time in 4 wt% NaCl solution. The OCP behavior of the graphene incorporated coating was very similar to the PVB/PPyCB20 composite coating. The OCP shifted from -0.12 V vs. SCE to -0.26 V vs. SCE during the initial 5 h immersion, followed by an increase to a value of -0.16 V vs. SCE at the end of the measurement. Such an OCP pattern suggested that corrosion inhibition took place through the healing of defects on the PVB/PPyCB20/Gr5 coating. EIS was performed to study the electrochemical responses of PVB/PPyCB20/Gr5 coating on steel substrate. The Bode plot measured during the initial time of immersion, shown in Figure 5.11b, exhibited two regions of distinct electrochemical responses. The first time constant at high frequencies (10^2 Hz to 10^4 Hz) was governed by the coating response. At low frequency, the kinetics of a charge-transfer process governed the response as manifested by the second time constant which indicated the electron transfer reaction between metal and conducting polymer. Such an electrochemical reaction (metal oxidation) at the metal-PVB/ PPyCB20/Gr5 coating interface resulted in a low frequency impedance in the Bode plot (Figure

5.11b). These results indicated that PVB/PPyCB20/Gr5 coatings were more prone to defects due to weak interaction of Gr with PVB matrix and metal. However, prolonged immersion in 4 wt% NaCl solution increased impedance especially in the low frequency region by an order of 10^2. Since low frequency impedance is directly related to the protective efficiency of the coating [52], it can be suggested that conducting particles, PPyCB pigment and graphene platelets, induced an inhibition to corrosion processes on the metal surface. These changes are more obvious in Bode phase angle plot shown in Figure 5.11b. The first time constant attributed to the coating response changed to a resistive plateau at 90° which indicated that the coating barrier effect enhanced with immersion time. The coating barrier is related to the resistance of the electrolyte in pores, cracks and defects on the coating [42]. Therefore, the behavior observed in the

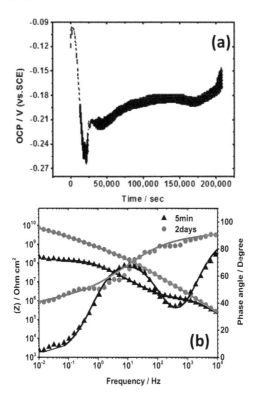

Figure 5.11 Electrochemical analysis conducted on graphene incorporated PVB/PPyCB20 composite coating during immersion in 4 wt% NaCl solution: (a) OCP vs. time of immersion, (b) Bode plots measured after 5 min and 2 d immersion, solid lines show the fit results.

PVB/PPyCB20/Gr5 coating can be considered to be caused by the passive layer deposition on the coating defects, which prevented the direct ingress of electrolyte down to the metal surface. It was more obvious in the second time constant, attributed to the charge transfer reactions on the metal/polymer interface, which completely vanished after 2 d immersion. In addition, the low frequency phase angle increased to 45°. These changes suggested that corrosion attack was effectively stopped by the passive layer deposition on the metal surface. Changes in the coating and on the metal surface were studied by fitting the EIS plots with an equivalent circuit shown in Figure 5.6 and the fitting data are displayed in Table 5.6. R_{ct}, corresponding to resistance to charge transfer process on the metal surface, increased significantly with immersion time which indicated the diminished corrosion process on the metal surface by the deposition of passive layer. The deposition of protective layers could be confirmed further from the deposited layer capacitance (C_{dl}):

$$C_{dl} = \varepsilon\varepsilon_o(At^{-1})$$

where ε and ε_o represent dielectric constant of deposition and the permittivity of free space (8.9×10^{-14} F cm^{-1}) respectively, A is the area of the exposed metal surface and t is the deposited layer thickness [53]. As Q_{dl} was used instead of pure capacitance, the decrease in Y_0 constant of Q_{dl} with immersion time indicated an increase in the thickness of deposited layers. The Y_0 values also indicate the area of metal surface exposed for the corrosion process [42,43], which decreased with immersion time. Such deposition of layers on the metal surface can be attributed to an increased n_2 value. The exponential term of the constant phase element n also measures the surface inhomogeneity; the lower is its value, the higher is the surface roughening of the metal/alloy and vice versa [54]. Therefore, the increased R_{ct}, n_1 and decreased Y_0 constant of Q_{dl} (Table 5.6) confirmed the formation of passive layers at the defect areas. As per the scheme shown in Figure 5.7, corrosion protection of the coating was controlled by polypyrrole's ability to capture and transport electrons from the metal surface. Since both conducting polymer and graphene platelets possess conjugated π systems, an electronic interaction via π–π stacking can be expected to occur [34,35]. It was observed in the PVB/PPyCB20/Gr5 coating that graphene promoted the electron transfer for the PPy redox processes and, thus, enhanced the protective passive layer formation on metal surface. Furthermore, it has been reported that graphene is an excellent barrier to oxygen and water diffusion [28-35]. The diminished water permeability of PVB/PPyCB20 coating with the addition of graphene resulted a decreased Y0 value of Qc in 2 d immersion. It was attributed further to an increase of coating resistance (R_c) (Table 5.6) as well. Therefore, it can be suggested from these results that the reinforced effect of electrical conductivity along with the reduced oxygen and water permeability by graphene incorporation enhanced the protective behavior of PVB/PPyCB20 coatings.

Table 5.6 EIS parameters of PVB/PPyCB20/Gr5 composite coatings at different time of immersion in 4 wt% NaCl solution

Time	R_s[a]	R_c[b]	Q_c			R_{ct}[e]	Q_{dl}	
	(Ω cm^2)	(Ω cm^2)	n_1[c]	Y_0[d] (Ω^{-1} cm^2)		(Ω cm^2)	n_2[f]	Y_0[g] (Ω^{-1} cm^2)
10 min	760	2.201e6	1	3.680e-10		10.01e6	0.556	22.76e-9
2 d	1860	6.915e9	1	40.56e-12		6.50e9	0.999	0.139e-9

[a] 3.6% probable error. [b] 8% probable error. [c] 1.5% probable error. [d] 14% probable error. [e] 10% probable error.
[f] 1.2% probable error. [g] 9% probable error.

Polarization measurements on the PVB/PPyCB20/Gr5 coating were conducted after 1 h and 1 d immersion in 4 wt% NaCl solution (Figure 5.12). E_{corr} values obtained from the polarization curves were

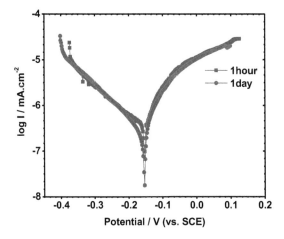

Figure 5.12 Polarization curves of graphene incorporated PVB/PPyCB20 composite coating after 1 h and 1 d immersion in 4 wt% NaCl solution.

-0.154 V vs. SCE and -0.156 V vs. SCE respectively for 1 h and 1 d exposed samples (Table 5.7). It indicated that the healing of defects by

Table 5.7 List of electrochemical parameters obtained from Tafel plots of PVB/PPyCB20/Gr5 composite coating after 1 h and 1 d immersion in 4 wt% NaCl solution

Time	E_{corr} (V vs. SCE)	I_{corr} (mA cm^{-2})	β_c (mV per decade)
1 h	−0.154	3.91×10^{-7}	144
1 d	−0.156	4.38×10^{-7}	151

the deposited passive layers helped the coating to maintain its protective performance. It is also seen that the cathodic branches of the polarization curves displayed a typical Tafel behavior whereas anodic polarization curves were deviated from the linear Tafel behavior over the complete applied potential range. Therefore, Tafel constants (I_{corr} and β_c) given in Table 5.7 were calculated from the extrapolation of the cathodic branch of the polarization curves. The curvature of the anodic branch was attributed to the deposition of passive layers on the metal surface. The protection offered by the deposited passive layer on the metal surface could be confirmed from the increase of β_c value after 1 d exposure in sodium chloride solution. The corrosion protection performance of PPyCB composite coatings resulted due to synergistic effect of electrical conductivity and redox processes of PPy. The presence of graphene particle in the composite improved the electrical conductivity of the coating, which, in turn, enhanced the redox processes of PPy as well. This probably attributed to the unexpected increase of Icorr of PVB/PPyCB20/Gr5 coating observed after 1 d immersion in sodium chloride solution. In addition, uniform distribution of graphene platelets enhanced the diffusion paths of water and ions through the coating, thus, effectively prevented its availability on the metal surface for the corrosion processes.

5.4 Conclusions

PPyCB pigment, up to a concentration of 20 wt%, was successfully incorporated into PVB coating formulation without affecting the barrier effect and adhesion properties of organic coatings. PPy imparted a self-healing character to PVB/PPyCB coatings that shifted the OCP of the metal from a dissolution potential to a noble potential on immersion in aggressive corrosive media. It was also reflected in the polarization curves of the PVB/PPyCB20 composite coatings with an anodic shift of E_{corr}. A synergistic effect of the redox properties and diffusion barrier nature of PPy resulted in the corrosion resistance performance of the PVB/PPyCB coatings. SEM and Raman spectra of the surface of intentionally made scratches on the coatings exhibited formation of passive layers due to the interaction between released dopant from organic sulfonic acid doped PPy and iron oxides. The deposited passive layer along with the reduced PPy hindered the diffusion of oxygen and water down to the metal surface, which diminished the corrosion processes. This was attributed to an increased R_{ct}, significant for 20% PPyCB composite coatings, with immersion time. Incorporation of conducting graphene particles in the PVB/PPyCB20 composite coatings enhanced the process of passive layer deposition on the metal surface, which hindered electron transfer between metal and polymer along with preventing water and gas from reaching the metal surface, thus, preventing metal from further corrosion. It can be concluded from these results that the use of conducting composite particles such as PPyCB as corrosion inhibiting pigments imparts a high degree of self-healing protective nature to the organic coatings without sacrificing the inherent coating properties.

Main Findings

Self-healing polyvinyl butyral (PVB) based organic coating formulations were prepared by incorporating polypyrrole-carbon black (PPyCB) composite as an inhibiting pigment. The redox properties and diffusion barrier nature of PPy imparted self-healing to the PVB/PPyCB composite coatings in aggressive environments. PPy induced the formation of a stable passive layer on the metal surface through the interaction of released dopant, from the organic sulfonic acid doped PPy with metal iron oxide. SEM images and Raman spectroscopy confirmed the formation of a protective passive layer on the metal surface. Furthermore, reduced PPy hindered the diffusion of water and oxygen through the coating. The addition of more conducting particles like graphene further enhanced the protective nature of the PVB/ PPyCB composite coatings. This work demonstrates a possible application of conducting particles in enhancing the protective nature of organic coatings widely used in industry.

Acknowledgement

This work has been published earlier in RSC Advances (2016) 6:43237-43249. The work has been reproduced here with permission from Royal Society of Chemistry.

References

1. Tian, Z., Yu, H., Wang, L., Saleem, M., Ren, F., Ren, P., Chen, Y., Sun, R., Sun, Y., and Huang, L. (2014) Recent progress in the preparation of polyaniline nanostructures and their applications in anticorrosive coatings. *RSC Advances*, **4**, 28195-28208.
2. Qi, K., Qiu, Y., Chen, Z., and Guo, X. (2012) Corrosion of conductive polypyrrole: Effects of environmental factors, electrochemical stimulation, and doping anions. *Corrosion Science*, **60**, 50-58.
3. Xing, C., Zhang, Z., Yu, L., Zhang, L., and Bowmaker, G. A. (2014) Electrochemical corrosion behavior of carbon steel coated by polyaniline copolymers micro/nanostructures. *RSC Advances*, **4**, 32718-32725.
4. Yan, M., Vetter, C. A., and Gelling V. J. (2013) Corrosion inhibition performance of polypyrrole Al flake composite coatings for Al alloys. *Corrosion Science*, **70**, 37-45.
5. Wessling, B. (1994) Passivation of metals by coating with polyaniline: corrosion potential shift and morphological changes. *Advanced Materials*, **6**, 226-228.
6. Lu, W. K., Elsenbaumer, R. L., and Wessling, B. (1995) Corrosion protection of mild steel by coatings containing polyaniline. *Synthetic Metals*, **71**, 2163-2166.
7. Paliwoda-Porebska, G., Stratman, M., Rohwerder, M., Potje-Kamloth, K., Lu, Y., Pich, A. Z., and Adler, H.-J. (2005) On the development of polypyrrole coatings with self-healing properties for iron corrosion protection. *Corrosion Science*, **47**, 3216-3233.
8. Iroh, J. O., and Su, W. (2000) Corrosion performance of polypyrrole coating applied to low carbon steel by an electrochemical process. *Electrochimica Acta*, **46**, 15-24.
9. Nguyen, T., Le, T., Garcia, B., Deslouis, C., and Le, X. Q. (2001) Corrosion protection and conducting polymers: polypyrrole films on iron. *Electrochimica Acta*, **46**, 4259-4272.
10. Kowalski, D., Ueda, M., and Ohtsuka, T. (2010) Self-healing ion-permselective conducting polymer coating. *Journal of Materials Chemistry*, **20**, 7630-7633.
11. Liu, A. S., Bezerra, M. C., and Cho, L. Y. (2009) Electrodeposition of polypyrrole films on aluminium surfaces from p-toluene sulfonic

acid medium. *Materials Research*, **12**, 503-507.
12. Balaska, A. C., Kartsonakis, I. A., Kordas, G., Cabral, A. M., and Morais, P. J. (2011) Influence of the doping agent on the corrosion protection properties of polypyrrole grown on aluminium alloy 2024\T3. *Progress in Organic Coatings*, **71**, 181-187.
13. Hien, N. T. L., Garcia, B., Pailleret, A., and Deslouis, C. (2005) Role of doping ions in the corrosion protection of iron by polypyrrole films. *Electrochimica Acta*, **50**, 1747-1755.
14. Krstajic, N. V., Grgur, B. N., Jovanovic, S. M., and Vojnovic, M. V. (1997) Corrosion protection of mild steel by polypyrrole coatings in acid sulfate solution. *Electrochimica Acta*, **42**, 1685-1691.
15. Armelin, E., Pla, R., Liesa, F., Ramis, X., Iribarren, J. I., and Aleman, C. (2008) Corrosion protection with polyaniline and polypyrrole as anticorrosive additives for epoxy paint. *Corrosion Science*, **50**, 721-728.
16. Bae, W. J., Kim, K. H., and Jo, W. H. (2005) A water-soluble and self-doped conducting polypyrrole graft copolymer. *Macromolecules*, **38**, 1044-1047.
17. Radhakrishnan, S., Siju, S. R., Mahanta, D., Patil, S., and Madras, G. (2009) Conducting polyaniline–nano-TiO$_2$ composites for smart corrosion resistant coating. *Electrochimica Acta*, **54**, 1249-1254.
18. Mahmoudian, M. R., Alias, Y., Basirun, W. J., and Yousefi, R. (2013) Synthesis and characterization of zinc/polypyrrole nanotube as a protective pigment in organic coatings. *Metallurgical and Materials Transactions A*, **44A**, 3353-3363.
19. Kraljic, M., Mandic, Z., and Duic, L. (2003) Inhibition of steel corrosion by polyaniline coatings. *Corrosion Science*, **45**, 181-198.
20. Pereira da Siva, J. E., Cordoba de Torresi, S. I., and Torresi, R. M. (2007) Polyaniline/poly(methylmethacrylate) blends for corrosion protection: The effect of passivating dopants on different metals. *Progress in Organic Coatings*, **58**, 33-39.
21. Radhakrishnan, S., Sonawane, N., and Siju, C. R. (2009) Epoxy powder coatings containing polyaniline for enhanced corrosion protection. *Progress in Organic Coatings*, **64**, 383-386.
22. Shao, Y., Huang, H., Zhang, T., Meng, G., and Wang, F. (2009) Corrosion protection of Mg–5Li alloy with epoxy coatings containing polyaniline. *Corrosion Science*, **51**, 2906-2915.
23. Ruhi, G., Modi, O. P., and Dhawan, S. K. (2015) Chitosan-polypyrrole-SiO$_2$ composite coatings with advanced anticorrosion properties. *Synthetic Metals*, **200**, 24-39.
24. Selvaraj, M., Palraj, S., Maruthan, K., Rajagopal, G., and Venkatachari, G. (2010) Synthesis and characterization of polypyrrole composites for corrosion protection of steel. *Journal of Applied Polymer Science*, **116**, 1524-1537.
25. Tong, Y., Bohm, S., and Song M. (2014) Graphene based materials

and their composites as coatings. *Journal of Nanomedicine and Nanotechnology*, **1**, 1003.
26. Singh, K., Ohlan, A., and Dhawan, S. K. (2012) Polymer-graphene nanocomposites: preparation, characterization, properties, and applications. In: *Nanocomposites - New Trends and Developments*, Ebrahimi, F. (ed.), InTech, Croatia, pp. 37-72.
27. Chaudhry, A. U., Mittal, V., and Mishra, B. (2015) Inhibition and promotion of electrochemical reactions by graphene in organic coatings. *RSC Advances*, **5**, 80365-80368.
28. Chang, K. C., Hsu, M. H., Lu, H. I., Lai, M.C., Liu, P. J., Hsu, C. H., Ji, W. F., Chuang, T. L., Wei, Y., Yeh, J. M., and Liu, W. R. (2014) Synergistic effects of hydrophobicity and gas barrier properties on the anticorrosion property of PMMA nanocomposite coatings embedded with graphene nanosheets. *Carbon*, **66**, 144-153.
29. Okafor, P. A., Singh-Beemat, J., and Iroh, J. O. (2015) Thermomechanical and corrosion inhibition properties of graphene/epoxy ester–siloxane–urea hybrid polymer nanocomposites. *Progress in Organic Coatings*, **88**, 237-244.
30. Li, Y., Yang, Z., Qiu, H., Dai, Y., Zheng, Q., Li, J., and Yang, J. (2014) Self-aligned graphene as anticorrosive barrier in waterborne pol-yurethane composite coatings. *Journal of Materials Chemistry A*, **2**, 14139-14145.
31. Mayavan, S., Siva, T., and Sathiyanarayanan, S. (2013) Graphene ink as a corrosion inhibiting blanket for iron in an aggressive chloride environment. *RSC Advances*, **3**, 24868-24871.
32. Sreevatsa, S., Banerjee, A., and Haim G. (2009) Graphene as a permeable ionic barrier graphene composites and gels. *ECS Transactions*, **19**, 259-264.
33. Chang, C.-H., Huang, T.-C., Peng, C.-W., Yeh, T.-C., Lu, H.-I., Hung, W.-I., Weng, C.-J., Yang, T.-I., and Yeh, J.-M. (2012) Novel anticorrosion coatings prepared from polyaniline/graphene composites. *Carbon*, **50**, 5044-5051.
34. Kirkland, N. T., Schiller, T., Medhekar, N., and Birbilis, N. (2012) Exploring graphene as a corrosion protection barrier. *Corrosion Science*, **56**, 1-4.
35. Merisalu, M., Kahro, T., Kozlova, J., Niilisk, A., Nikolajev, A., Marandi, M., Floren, A., Alles, H., and Sammelselg, V. (2015) Graphene-polypyrrole thin hybrid corrosion resistant coatings for copper. *Synthetic Metals*, **200**, 16-23.
36. Shi, Z., Liu, M., and Atrens, A. (2010) Measurement of the corrosion rate of magnesium alloys using Tafel extrapolation. *Corrosion Science*, **52**, 579-588.
37. McCafferty, E. (2005) Validation of corrosion rates measured by the Tafel extrapolation method. *Corrosion Science*, **47**, 3202-3215.
38. Amin, M. A., Abd El Rehim, S. S., and Abdel-Fatah, H. T. M. (2010) Te-

sting validity of the Tafel extrapolation method for monitoring corrosion of cold rolled steel in HCl solutions – Experimental and theoretical studies. *Corrosion Science*, **52**, 140-151.
39. Maranhao, S. L. A., Guedes, I. C., Anaissi, F. J., Toma, H. E., and Aoki, I. V. (2006) Electrochemical and corrosion studies of poly(nickel-tetraaminophthalocyanine) on carbon steel. *Electrochimica Acta*, **52**, 519-526.
40. Carneiro, J., Tedim, J., Fernandes, S. C. M., Freire, C. S. R., Silvestre, A. J. D., Gandini, A., Ferreira, M. G. S., and Zheludkevich, M. L. (2012) Chitosan-based self-healing protective coatings doped with cerium nitrate for corrosion protection of aluminum alloy. *Progress in Organic Coatings*, **75**, 8-13.
41. Akid, R., Gobara, M., and Wang, H. (2011) Corrosion protection performance of novel hybrid polyaniline/sol–gel coatings on an aluminium 2024 alloy in neutral, alkaline and acidic solutions. *Electrochimica Acta*, **56**, 2483-2492.
42. Zheludkevich, M. L., Serra, R., Montemor, M. F., Yasakau, K. A., Salvado, I. M. M., and Ferreira, M. G. S. (2005) Nanostructured sol–gel coatings doped with cerium nitrate as pre-treatments for AA2024-T3: Corrosion protection performance. *Electrochimica Acta*, **51**, 208-217.
43. Latnikova, A., Grigoriev, D., Schenderlein, M., Mohwald, H., and Shchukin, D. (2012) A new approach towards "active" self-healing coatings: exploitation of microgels. *Soft Matter*, **8**, 10837-10844.
44. Conceicao, T. F., Scharnagl, N., Blawert, C., Dietzel, W., and Kainer, K. U. (2010) Corrosion protection of magnesium alloy AZ31 sheets by spin coating process with poly(ether imide). *Corrosion Science*, **52**, 2066-2079.
45. Lopez Cascales, J. J., Fernandez, A. J., and Otero, T. F. (2003) Characterization of the reduced and oxidized polypyrrole/water Interface: a molecular dynamics simulation study. *Journal of Physical Chemistry B*, **107**, 9339-9343.
46. Levine, K. L. (2013) Nanocomposite PPy coatings for Al alloys corrosion protection. In: *Polymer Nanocomposite Coatings*, Mittal, V. (ed.), CRC Press, USA, pp. 277-296.
47. Reut, J., Opik, A., and Idla K. (1999) Corrosion behavior of polypyrrole coated mild steel. *Synthetic Metals*, **102**, 1392-1393.
48. Yeh, J. M., Chen, C. L., Chen, Y. C., Ma, C. Y., Lee K. R., Wei, Y., and Li, S. (2002) Enhancement of corrosion protection effect of poly(o-ethoxyaniline) via the formation of poly(o-ethoxyaniline)–clay nanocomposite materials. *Polymer*, **43**, 2729-2736.
49. Olad, A., Rashidzadeh, A., and Amini, M. (2013) Preparation of polypyrrole nanocomposites with organophilic and hydrophilic montmorillonite and investigation of their corrosion protection on iron. *Advances in Polymer Technology*, **32**(3), 21337.

50. Ruhi, G., Bhandari, H., and Dhawan, S. K. (2015) Corrosion resistant polypyrrole/flyash composite coatings designed for mild steel substrate. *American Journal of Polymer Science*, **5**(1A), 18-27.
51. Meroufel, A., Deslouis, C., and Touzain, S. (2008) Electrochemical and anticorrosion performances of zinc-rich and polyaniline powder coatings. *Electrochimica Acta*, **53**, 2331-2338.
52. Sugama, T., and Cook, M. (2000) Poly(itaconic acid)-modified chitosan coatings for mitigating corrosion of aluminum substrates. *Progress in Organic Coatings*, **38**, 79-87.
53. López, D. A., Simison, S. N., and de Sánchez, S. R. (2003) The influence of steel microstructure on CO_2 corrosion. EIS studies on the inhibition efficiency of benzimidazole. *Electrochimica Acta*, **48**(7), 845-854.
54. Chongdar, S., Gunasekaran, G., and Kumar, P. (2005) Corrosion inhibition of mild steel by aerobic biofilm. *Electrochimica Acta*, **50**(24), 4655-4665.

6

UV Degradation of Polymer Coatings

6.1 Introduction

Polymeric coatings are developed for meeting various needs, such as corrosion protection, moisture and gas barrier, maintenance of equipment and surfaces, etc. The protective polymeric coatings are of substantial importance in present-day technologies, and extensive research efforts have been carried out for the advancement of such coatings with enhanced environmental stability. Unlike the architectural coatings, which are largely used in amiable weather conditions, the protective coatings are usually applied in challenging and extreme environments. As an illustration, the pipelines with surface coatings, employed for the transportation of oil, gas or chemicals over longer distances, are subjected to severe hot, cold, arid and humid conditions.

Organic coatings are extensively employed for preventing the corrosion of the metallic structures due to their effective properties, low manufacturing cost, versatility, ease of application and aesthetic features [1]. Different physical processes as well as chemical reactions take place when the synthetic or natural materials based coatings are exposed to outdoor conditions for prolonged periods of time. Such physical processes and chemical reactions are collectively termed as weathering [2]. The exterior durability of an organic coating is described as the resistance offered by the material to the unwanted effects induced by the natural environment to which the coating is exposed to during its service life [3]. The longer a coating is capable of avoiding failure because of its optimal weathering resistance, the better is its durability and dimensional stability. Service lifetime is defined as the time that a coating can last till the failure occurs. The service lifetime of a polymer coating is determined by the coating characteristics as well as the service conditions in which the coating is located. Overall, the development of coatings with superior weathering resistance, along with other property specifications, is an important research area.

Haleema Saleem and Vikas Mittal, Khalifa University of Science and Technology, Abu Dhabi, UAE
© 2021 Central West Publishing, Australia

Various studies have been carried out to understand the degradation behavior of the polymer coatings in the past several years. The ultraviolet (UV) radiation, O_2 and H_2O are the three discriminating factors responsible for the degradation of coatings during weathering [4]. When the aforementioned factors are united with several other environmental variables like wind and seasonal periodicity, the achievement of weather resistant coatings becomes even more challenging.

The UV light is an electromagnetic radiation having wavelength between 10-400 nm and energy ranging from 3-124 eV [5]. As per the ISO solar irradiance standard (ISO 21348), the electromagnetic spectrum of UV is subdivided into the following three major groups [5,6]:

- Ultraviolet A (UVA): These radiations have wavelength in the range 320-400 nm. Almost 99% of the total UV light reaching the earth's surface is UVA. It is accountable for several photo-sensitivity reactions, and it can enhance the adverse effects of ultraviolet B radiations.
- Ultraviolet B (UVB): Nearly 1% of the total UV light reaching the earth's surface is UVB. These radiations have wavelength in the range 290-320 nm. UVB generates several detrimental photo-chemical reactions.
- Ultraviolet C (UVC): These radiations have wavelength in the range 200-290 nm. Usually, UVC cannot reach the earth's surface due to the blockage by the ozone layer.

For majority of the synthetic polymers, the most substantial degradation mechanisms are linked to the absorption of UV light having energy between 300-450 kJ/mol [7]. When the quantity of energy absorbed by the polymer outstrips the polymer bond energy, the UV degradation takes place [7]. The rate of polymer degradation depends on various factors like the exposure location, temperature, nature of the substrate, polymer material, etc. The degradation of the polymeric coatings can be established in the form of swelling, color variation, crosslinking, oxidation, water absorption and dissolution [6]. Further, at elevated temperatures, certain gas species may be generated from the coatings, thereby, varying the glass transition temperature, molecular weight, gloss and density of the polymer matrix in the coatings. These factors generally lead to an increment in the brittleness and porosity of the polymeric coatings [8,9]. The photo-degradation of the crosslinked (thermosetting) polymer coatings generally also causes variations in the crosslink density that can change the glass transition temperature. Thus, the recognition of photo-degradation is connected to realizing the physical and chemical effects as well as the relations between them. The photo-degradation processes in the top layers of the polymeric coatings are spatially non-uniform in nature because of the nature of UV absorptivity of the polymeric materials and ingress of water and oxygen. Prominent examples of the physical properties of the polymer coatings affected by the photo-degradation processes include oxygen permeability [10], optical properties (color, UV-vis absorption) [11,12], water sorption [13], gloss [14,15], fracture energy [16], surface tension [14,15], hardness [17], internal stresses [18], elastic modulus [19], crosslink density [18] and glass transition temperature [20]. Conventional failure modes are related to the coating appearance as well as its mechanical integrity (delamination and cracking failure). The failure associated with the crack formation is due to the fact that the coatings turn brittle at the time of degradation and both quick fatigue failure and brittle failure may take place [21]. Various degradation factors affect the failure modes in distinct ways and each failure mode might exhibit a diversified dependence on particular coating feature, e.g., the layer thickness [16].

Due to weathering, the service life of polymers in the open air applications becomes limited. The service life of the polymeric coatings can be considerably lengthened, either by the surface treatment to screen the adverse UV radiations or by the incorporation of light stabilizers such as hindered amine light stabilizer (HALS), UV absorbers, etc. [22]. For the development of polymeric coatings with improved properties, characterization of the long-term degradation is also a fundamental necessity, and various techniques have been employed in this respect. For instance, several techniques address the macroscopic properties of the system such as weight, mechanical integrity, loss of gloss and variation in contact angle [23]. These techniques provide information associated with the material performance [24], however, these do not give insights about the atomic scale degradation. On the other hand, Fourier transform infrared (FTIR) spectroscopy, scanning electron microscopy (SEM), electron spin resonance (ESR), Raman spectroscopy, nuclear magnetic resonance (NMR) and positron annihilation spectroscopy (PAS) are extensively used for analyzing the microscopic as well as atomic scale degradation [25,26]. Out of these, PAS and ESR are specifically employed for identifying the early stages of coating degradation.

An improved understanding of the degradation behavior supports the forecasting of the performance of the polymer coatings, along with tuning of the coatings formulations for enhancing their durability. In this respect, the photo-degradation of diverse polymer coatings (e.g., acrylic, epoxy, polyurethane (PU), silicone and polyester coatings) has been discussed in detail in the current chapter.

6.2 Mechanism of UV Degradation

The UV degradation of the polymer materials is generally induced by a complex series of chemical reactions, which are brought about by the UV light absorption. The physical properties of the polymer deteriorate as a result of the UV degradation, which is one of the most common issues seen in the polymeric coatings exposed to sunlight. The continuous exposure to UV light causes more severe damage than the intermittent exposure. Based on the changes occurring in the polymers under the effect of UV radiation exposure, the synthetic polymers might be divided into two groups. The first group consists of the polymers that discolor very quickly when exposed to UV radiation, however, maintain the physical properties for continued periods of irradiation. Due to the UV radiation exposure, changes occur in the chemical structure, thereby, activating the chromophoric groups. Nevertheless, the scission of polymer backbone does not occur in these type of polymers. Polyacrylonitrile (PAN) and polyvinylchloride (PVC) are examples of these types of polymers [27]. The second type of polymers are those which become brittle during the exposure of UV radiation. This is due to the breaking of the main chains as well as photo-induced crystallization [6]. Polymers such as polyethylene, polystyrene and polypropylene are the examples of this category. The UV light of high intensity causes the generation of free radicals on the surface of the polymer [28]. Hence, this initiates the crosslinking reactions for the additional polymerization, oxidation or both [5]. These radicals can easily react with the oxygen present in the air, thereby, enabling auto-accelerating photo-oxidation. Environmental conditions like temperature, humidity, acidity/basicity, pollutants and oxygen remarkably enhance the UV degradation level [29]. The presence of water is very crucial because of its direct participation during the degradation process. The water molecules may also lead to the matrix plasticization that can change the polymer coating's effective glass transition temperature, solubility and diffusion coefficient of additional degradation agents like oxygen [30].

Similar to the radical polymerization chemistry, the photo-degradation process is defined in terms of initiation, propagation and termination reactions [31]. The initiation process includes a photolytic scission that takes place after the UV absorption by the polymer molecules, thus, causing the chain scission into two radicals. The polymer radical is effortlessly oxidized to obtain a peroxy radical, which can abstract a hydrogen atom deriving from a different polymer fragment in the coating to obtain hydroperoxide as well as additional polymer radical. The hydroperoxide decomposes to form a polymer oxy-radical as well as a hydroxy radical under the influence of temperature and photons. Both radical species take part in the hydrogen abstraction processes and form new polymer radicals, which can again take part in the oxidation reactions. Further, the polymer oxy-radical present in the polymer chains can cause chain scission. Finally, the terminal reactions occur via radical recombination process.

As mentioned earlier, the durability of the polymer coatings can be enhanced by the incorporation of stabilizers in the formulations. These stabilizers have the ability to target the initiation as well as propagation reactions during the photo-degradation process. The initiation reactions can be limited by avoiding a part of the incident UV radiation from being absorbed by the polymer. This can be accomplished by the reflection and absorption of the incident radiation as well as the consequent dissipation of its energy as heat and long wavelength radiation [32]. Apart from avoiding the absorption process, the initiation process can also be avoided after the development of an absorption event, in case the excited chromophore energy is moved to a stabilizer molecule before the chain breaking occurs. This method is known as the quenching of the excited states and can be obtained by the incorporation of metal chelate compounds to the polymeric coatings [33]. Not every UV photon can be counterbalanced by the stabilization mechanisms, thus, the initiation might take place partially. Due to this reason, radical scavengers might be incorporated in the coatings formulations for partially avoiding the propagation processes. The most widely employed radical scavenger is HALS, which is based on nitroxy radicals (R-NO•). The nitroxy radicals recombine with the polymer radicals to generate alkyloxyamine. The selection of additives that constitute the stabilizing system has to be tuned on the basis of the photo-degradation mechanism of the particular polymer needed to be stabilized. Hence, an excellent knowledge of the photo-degradation mechanisms of the polymeric materials is essential for effective stabilization.

6.3 Different Polymer Coatings Systems

Polymer resins are the starting component of all coating formulations and inevitably guide the fundamental properties of the coatings systems. Unlike the architectural coatings where the water borne coatings systems dominate, the protective coatings dominantly rely on the solvent borne systems, as these facilitate the usage of highly durable and high T_g resins. The UV degradation performance of different polymeric coatings based on acrylics, epoxies, PUs, silicones and polyesters is discussed in the following sections.

6.3.1 PU Coatings

PUs are employed in an extensive variety of technical as well as commercial applications, due to their chemical resistance, good mechanical properties, processability and high tensile strength [34]. PU coatings are commonly utilized for the protection of materials such as metals, wood and plastics because of their color retention, excellent gloss and dimensional stability [35]. One of the important drawbacks of PU based coatings is severe light sensitivity, specifically UV light. This limits the application of PU materials as the surface coatings for external utilization. When exposed to high energy UV radiations, PU goes through remarkable structural variation, which leads to the deterioration in its physical properties [36,37]. PU materials prepared using aromatic isocyanates exhibit yellowing, when polymer coating is exposed to UV light, due to the oxidation process occurring in the polymer backbone. The stabilization against high energy radiations can be achieved by the addition of photo-stabilizers and anti-oxidants [36]. Zinc oxide (ZnO) and titanium dioxide (TiO_2) have been observed to be effective additives that promote good photo-stabilization effect in PUs.

Yang et al. [4] analyzed the degradation of epoxy primer with high gloss PU topcoat coating system, which was exposed either in a prohesion chamber and a QUV chamber alternatively, or only in a QUV chamber. The atomic force microscopy (AFM) analysis confirmed the formation of micro-blisters on the coating surface after both exposures. During the QUV exposure, the blisters increased in size on enhancing the exposure time. However, in the case of prohesion/QUV alternative exposure, the blisters maintained their size throughout the time of exposure. The surface roughness (RMS) exhibited a steady enhancement with the time of QUV exposure, however, it was maintained constant during the prohesion/QUV exposure. The SEM analysis proved that the exposure to QUV was more destructive than the prohesion/QUV alternative exposure. In a dry/wet interchanging environment, the blisters with sub-micrometer to micrometer dimensions were formed on the coating surface.

He et al. [23] examined the photo-degradation of PU coatings by the electron spin resonance (ESR) method. The PU film specimens were subjected to various accelerated ageing conditions, which included narrow band UV irradiation of 340 and 313 nm, increased temperature without any irradiation and broadband irradiation from a xenon arc lamp. In addition, the effect of titania on the free radical generation was also investigated. It was observed that the radicals decayed in a period of 1 hour following the irradiation, in the presence of ambient oxygen. However, in the absence of oxygen, two long lifetimes of the order of 60 and 350 hours were detected. In another study, Singh et al. [38] reported a comprehensive yellowing of the urethane clear coatings based on castor oil and diphenylmethane diisocyanate (MDI) in the presence of sunlight. The clear aromatic PU coatings were attained by combining MDI and castor oil in 1:1 ratio at room temperature and subsequently applying on glass plate. To provide light stability and to restrain the yellowing in aromatic PU coatings, modification with stabilizers as well as their synergistic mixtures was suggested by Decker and Bandaikha [39].

6.3.2 Acrylic Coatings

The acrylic/methacrylic polymers have been widely employed in various industrial fields. In the formulation of surface coatings and paints, these polymers impart chemical stability, adhesion, mechanical properties and optical clarity [40]. The photo-stability of the aliphatic methacrylic and acrylic polymers is usually higher, when compared to polyolefins. Carbonyl ester groups present in these polymers are not precisely photo-chemically active. Further, the content of the trace impurities that can initiate the photo-degradation is generally low [41]. On the other hand, it was also proved that the photo-induced oxidation of the methacrylic and acrylic polymers is not autocatalytical, as seen in the case of polyolefins, however, it advances at a constant oxygen consumption rate [42].

Several studies have reported the weathering degradation behavior of the acrylic coatings [1,43]. Hu et al. [1] examined the ageing characteristics of the acrylic PU varnish coatings in two artificial weathering environments. The results confirmed that the Xenotest protocol had a remarkable influence on the gloss loss and thickness loss than the UV/condensation weathering exposure. From the FTIR analysis, it was demonstrated that the same degradation mechanism persisted in both weathering conditions.

The thermoset acrylic-melamine coatings are extensively utilized for exterior applications. These are generated by the reaction of acrylic polyol with alkylated melamine formaldehyde (MF) resin. Due to the fact that the reactions are reversible, the acrylic-melamine coatings are prone to degradation when exposed to a weathering environment. From the accelerated tests and outdoor exposures, it has been confirmed that the degradation of these coatings is highly affected by UV radiation, air pollutants and relative humidity [44,45]. Under the influence of UV light, these coatings go through chain scission reactions, thereby, generating amines as well as different carbonyl derivatives as degradation products [46,47].

Nguyen *et al.* [48] studied the effect of relative humidity (RH), ranged from 0-90%, on the moisture enhanced photolysis (MEP) of the moderately methylated melamine acrylic polymer coatings, which were exposed to UV light at 50 °C. Entire degradation under UV conditions at a specified relative humidity could be classified into four modes: the reactions occurring during the post-curing, dark hydrolysis at a specific RH, photolysis and MEP. It was noted that on increasing the RH, the rate and intensity of MEP also enhanced. Further, the increased degradation was described by a mechanism based on the hydrolysis-developed formaldehyde molecules which acted as chromophores for absorbing the UV light, thus, accelerating the photo-oxidation process.

During the photolysis as well as photo-oxidation of poly(methyl methacrylate) (PMMA) at 254 nm, comprehensive random chain breaking, followed by the generation of low molecular weight gas products and monomer molecules were observed [49,50]. The extent of scission was observed to be proportional to the UV radiation dose, and it was greater in inert conditions than air. The scission rate changed with the wavelength of radiation, attaining an ultimate value at 280 nm and reaching zero at wavelengths greater than 320 nm [51]. This phenomenon was described by assuming the effect of ketone or aldehyde groups, which are UV active at 280 nm. The PMMA photolysis occurred as a result of UV absorption, which promoted the de-esterification process due to hemolytic bond breaking. Chiantore *et al.* [52] also examined the photo-oxidative stability of the methacrylic as well as acrylic based polymers, with potential use as protective coatings for various substrates, under artificial solar light radiation. The polymers analyzed in the study included poly(ethyl acrylate), poly(methyl acrylate) (PMA), poly(butyl methacrylate) (PBMA) and poly(ethyl methacrylate). The methacrylate units were observed to be less reactive towards the oxidation process, when compared to the acrylate units. It was also noticed that the degradation of PBMA advanced through an entirely different mechanism, with considerable crosslinking as well as concurrent fragmentation reactions.

Decker *et al.* [53] achieved photo-degradation resistant materials by conserving the surface by using UV cured coatings with HALS radical scavenger as well as a UV absorber (phenyltriazine). The solvent-free UV curable polyurethane-acrylate (PUA) resin consisted of three components, namely aliphatic PU-acrylate telechelic oligomer, hexanediol diacrylate (reactive diluent) and bisacylphosphine oxide photo-initiator. Decker *et al.* [54] also studied the light stability of the water-based UV cured PUA coatings, tested in an accelerated QUV-A weatherometer. The IR spectroscopy was employed to analyze the fast polymerization of the acrylate double bonds at the time of severe illumination. UV curing process was scarcely altered by the incorporation of HALS radical scavengers as well as UV absorbers. The urethane unit (C-NH) was noted to be more sensitive to the photo-degradation process.

In another study, Larche *et al.* [43] predicted the service life of acrylic-melamine and acrylic-urethane coatings under UV light exposure. It was confirmed that the UV irradiation resulted in chain scission and crosslinking in the stabilized and un-stabilized coatings. The breaking of the ether and urethane bonds by UV light caused the generation of free radicals which acted as the source for crosslinking. In the presence of moisture, the photo-degradation of the acrylic-melamine coatings was partially enhanced. The presence of water increased the degradation of moderately methylated melamine acrylic coatings with the generation of formaldehyde by the hydrolysis process. The formaldehyde molecules absorbed UV radiation and dissociated to produce free radicals, followed by the abstraction of hydrogen in the melamine chains, thereby leading to the generation of amine as well as amide based products [48,55]. Also, the weathering degradation of UV cured acrylic coatings constituted chain scission as well as crosslinking. Nevertheless, because of the elaborate network structure, these polymeric coatings were observed to have superior weather resistant than the acrylic-melamine and acrylic-urethane coatings [52-54].

6.3.3 Epoxy Coatings

Epoxies represent an important class of protective polymeric coatings owing to adhesion as well as resistance to chemicals, corrosion, acids and hydrocarbons [56]. During the course of service of the epoxy coatings, the environmental factors like sunlight, atmosphere, temperature and humidity can cause degradation in their properties, thus, limiting their performance [57]. The discoloration and chalking of the epoxy coatings in the presence of UV is regarded as the dominant cause of concern. Several studies have examined the effect of UV light on the degradation behavior of amine cured diglycidyl ether of bisphenol A (DGEBA) epoxy resins [58,59]. It was proposed that the degradation in the chemical structure primarily resulted from the generation of carbonyl products obtained from phenoxy groups, development of amide functions linked to amine concentration and chain breaking processes. Rivaton *et al.* [60] reported that the photo-oxidation process of the phenoxy resins primarily involved the reactions of the aromatic ether units and breaking of CH_3-C bond of the iso-propylidene groups. It was proved that the UV light at the surface of the coating produced a strong oxidative and thermal degradation in the presence of oxygen [61]. The oxidative degradation on the surface of the epoxy coatings activated the microscopic physical defects as well as deformation, present at the molecular and atomic levels at the time of early ageing stages. With increasing the time of irradiation, these physical defects developed and eventually caused the failure of the coating system. Hence, it is critical to understand the changes in the microstructure (like pores, defects and free volume), specifically their impact on the water transportation behavior and anti-corrosion capability of the coatings, during the degradation process.

Fuwei *et al.* [62] examined the development of chemical functional groups, water barrier behavior and microstructure of polyamide-cured epoxy (DGEBA) coatings during UVA photo-oxidative ageing. During the early stages of ageing, reduction in the *S* parameter as well as the water uptake coefficient illustrated the generation of a highly compact structure caused by post-curing process (*S* parameter is described as the ratio of the central area to the total area in a Doppler broadening energy spectroscopy (DBES) spectrum). In the subsequent ageing process, the generation of carbonyl groups and molecular chain breaking was observed. From the electrochemical impedance spectroscopy (EIS) analysis, after 208 hours of UV irradiation, a new time constant was observed to develop at relatively high frequency (3.5×10^2 Hz). This revealed that a micro-porous layer developed near the DGEBA film surface.

Using the FTIR analysis, it is possible to predict the long-term performance of the materials [63-65]. In one such study, Gerlock et al. [66] employed the transmission FTIR analysis and hydroperoxide concentration behavior analysis for comparing the photo-oxidation protection of acrylic-melamine based clearcoats. Penon et al. [67] also analyzed the UV degradation behavior of pigmented epoxy coatings for different periods at elevated pressures (1-100 bar). Further, the dielectric sorption analysis (DSA) of the specimens was also carried out. The variation in the dynamics of the absorption characteristics of the degraded polymer resulted due to the enhanced hydrophilicity, porosity and crosslinking. The degradation at all pressures exhibited the desorption behavior, induced by the shrinking of the pore size and swelling of the polymeric coatings. With enhancing the pressure, an increase in the water sorption characteristics was observed in DSA, where a linear trend was observed till 50 bar.

6.3.4 Polyester Coatings

Polyester based coatings are superior candidates for the outdoor applications due to improved mechanical properties as well as outdoor durability. Santos et al. [68] examined the degradation performance of polyester and silicone polyester coatings, which were exposed to high UV conditions (two accelerated UV analyses and one natural atmosphere test). The analysis of the coating degradation was performed in accordance with ISO 4628 standard [69]. The coatings exhibited greater color variation and higher gloss loss after exposure of 24 days. It was observed that the organic pigments offered bright colors as well as higher gloss range, however, the inorganic pigments provided better resistance to UV. The polyesters based on isophthalic acid (IPA) have also been generally acknowledged for their weathering stability [70,71].

Adema et al. [72] also analyzed the artificial weathering of polyester-urethane coatings using FTIR spectroscopy and UV-vis spectroscopy. The reduction in the urethane crosslinks present in the polymer coatings was observed to take place faster and to a greater extent, when compared to the ester bond breaking. The results obtained from the chemical and optical characterization were used by the authors to propose a kinetic model for the ester bond photolysis, which contributed towards the assessment of the quantum effectiveness of the process.

6.3.5 Silicone Coatings

Silicones, generally termed as polysiloxanes, exhibit high UV resistance because of higher energy bonds and absence of conjugation. Mitra *et al.* [73] examined various dynamic mechanical and physical properties of aliphatic PU, alkyd modified PU, high performance aliphatic PU and cycloaliphatic epoxy modified polysiloxane based coatings before and after artificial weathering. Accelerated weathering test was carried out in a QUV chamber as per ASTM G 154 standard for about 30 days. During the artificial weathering test, the topcoat films were exposed cyclically to 313 nm UV-B radiation at 60 °C for 4 hours, followed by water condensation for 4 hours at 50 °C. Cycloaliphatic epoxy modified polysiloxane based coatings displayed nearly no loss of gloss. Overall, these coatings exhibited excellent weatherability as well as good chemical resistance due to comparatively inert Si-O backbone.

6.4 Summary and Outlook

The exposure to ultraviolet (UV) light brings about remarkable degradation in the polymeric coatings. The UV light generates photo-oxidative degradation processes, which cause scission of the polymer chains, thereby producing free radicals and lowering the molecular weight. This induces deterioration in the mechanical properties of the polymeric coatings. The degradation of the polymeric coatings is established in the form of dissolution, crosslinking, color variation, water absorption, oxidation and swelling. The service life of the polymeric coatings can be considerably enhanced, either by the surface treatment to screen the adverse UV radiation or by the incorporation of light stabilizers such as HALS radical scavengers, UV absorbers, etc.

References

1. Hu, J., Li, X., Gao, J., and Zhao, Q. (2009) Ageing behavior of acrylic

polyurethane varnish coating in artificial weathering environments. *Progress in Organic Coatings*, **65**(4), 504-509.
2. Guillet, J. E. (1972) Fundamental Processes in the UV degradation and stabilization of Polymers. *Pure and Applied Chemistry*, **30**(1-2), 135-144.
3. Johnson, B. W., and McIntyre, R. (1996) Analysis of test methods for UV durability predictions of polymer coatings. *Progress in Organic Coatings*, **27**(1), 95-106.
4. Yang, X. F., Tallman, D. E., Bierwagen, G. P., Croll, S. G., and Rohlik, S. (2002) Blistering and degradation of polyurethane coatings under different accelerated weathering tests. *Polymer Degradation and Stability*, **77**(1), 103-109.
5. *Handbook of UV Degradation and Stabilization*, Wypych, G. (ed.), ChemTec Publishing, USA (2010).
6. *Handbook of Polymer Degradation*, Hamid, S. H., Amin, M. B., and Maadhah, A. G. (eds.), Marcel Dekker, USA (1992).
7. Katangur, P., Patra, P. K., and Warner S. B. (2006) Nanostructured ultraviolet resistant polymer coatings. *Polymer Degradation and Stability*, **91**(10), 2437-2442.
8. Revie, R. W. (2008) *Corrosion and Corrosion Control: An Introduction to Corrosion Science and Engineering*, 4th edition, John Wiley & Sons, USA.
9. Ahmad, Z. (2006) Principles of Corrosion Engineering and Corrosion Control, Butterworth-Heinemann, UK.
10. Gardette, J.-L., Colin, A., Trivis, S., German, S., and Therias, S. (2014) Impact of photooxidative degradation on the oxygen permeability of poly(ethyleneterephthalate). *Polymer Degradation and Stability*, **103**, 35-41.
11. Croll, S. G., and Skaja, A. D. (2003) Quantitative spectroscopy to determine the effects of photodegradation on a model polyester-urethane coating. *Journal of Coatings Technology*, **75**, 85-93.
12. Skaja, A. D., and Croll, S. G. (2003) Quantitative ultraviolet spectroscopy in weathering of a model polyester-urethane coating. *Polymer Degradation and Stability*, **79**, 123-131.
13. Nichols, M., Boisseau, J., Pattison, L., Campbell, D., Quill, J., Zhang, J., Smith, D., Henderson, K., Seebergh, J., Berry, D., Misovski, T., and Peters, C. (2013) An improved accelerated weathering protocol to anticipate Florida exposure behavior of coatings. *Journal of Coatings Technology and Research*, **10**, 153-173.
14. Glockner, P., Ritter, H., Osterhold, M., Buhk, M., and Schlesing, W. (1999) Effect of weathering on physical properties of clearcoats. *Die Angewandte Makromolekulare Chemie*, **269**, 71-77.
15. Croll, S. G., Hinderliter, B. R., and Liu, S. (2006) Statistical approaches for predicting weathering degradation and service life. *Progress in Organic Coatings*, **55**, 75-87.

16. Nichols, M. E., Gerlock, J. L., Smith, C. A., and Darr, C. A. (1999) The effects of weathering on the mechanical performance of automotive paint systems. *Progress in Organic Coatings*, **35**, 153-159.
17. Larché, J. F., Bussière, P. O., Wong-Wah-Chung, P., and Gardette, J. L. (2012) Chemical structure evolution of acrylic-melamine thermoset upon photo-ageing. *European Polymer Journal*, **48**, 172-182.
18. Croll, S. G., Shi, X., and Fernando, B. M. D. (2008) The interplay of physical aging and degradation during weathering for two cross-linked coatings. *Progress in Organic Coatings*, **61**, 136-144.
19. Gu, X., Michaels, C. A., Drzal, P. L., Jasmin, J., Martin, D., Nguyen, T., and Martin, J. W. (2007) Probing photodegradation beneath the surface: a depth profiling study of UV-degraded polymeric coatings with microchemical imaging and nanoindentation. *Journal of Coatings Technology and Research*, **4**, 389-399.
20. Larche, J. F., Bussiere, P. O., and Gardette, J. L. (2011) Photo-oxidation of acrylic-urethane thermoset networks. Relating materials properties to changes of chemical structure. *Polymer Degradation and Stability*, **96**, 1438-1444.
21. Croll, S. G. (2013) Application and limitations of current understanding to model failure modes in coatings. *Journal of Coatings Technology and Research*, **10**, 15-27.
22. Decker, C., and Zahouily, K. (1999) Photodegradation and photooxidation of thermoset and UV-cured acrylate polymers. *Polymer Degradation and Stability*, **64**, 293-304.
23. He, Y., Yuan, J. P., Cao, H., Zhang, R., Jean, Y. C., and Sandreczki, T. C. (2001)Characterization of photo-degradation of a polyurethane coating system by electron spin resonance, *Progress in Organic Coatings*, **42**(1-2), 75-81.
24. *Characterization of Polymers*, Tong, H.-M., Kowalczyk, S. P., Saraf, R., and Chou N. J. (eds.), Butterworths Heinemann, USA (1994).
25. *Multidimensional Spectroscopy of Polymers: Vibrational, NMR, and Fluorescence Techniques*, Urban, M. W., and Provder, T. (eds.), American Chemical Society, USA (1995).
26. Cao, H., Zhang, R., Sundar, C. S., Yuan, J.-P., He, Y., Sandreczki, T. C., Jean, Y. C., and Nielsen, B. (1998) Degradation of polymer coating systems studied by positron annihilation spectroscopy. 1. UV irradiation effect. *Macromolecules*, **31**(19), 6627-6635.
27. Allen, N. S., and Edge, M. (1993) *Fundamentals of Polymer Degradation and Stabilization*, Springer, Netherlands.
28. Asmatulu, R., Claus, R. O., Mecham, J. B., and Corcoran, S. G. (2007) Nanotechnology-associated coatings for aircrafts, *Journal of Materials Science*, **43**, 415-422.
29. Fernando, R. H. (2009) Nanocomposite and nanostructured coatings: recent advancements, Nanotechnology Applications in Coating. *ACS Symposium Series*, **1008**, 2-21.

30. Asmatulu, R., and Revuri, S. (2008) Synthesis and Characterization of Nanocomposite Coatings for the Prevention of Metal Surfaces. *Proceedings of SAMPE Fall Technical Conference*, USA, pp. 1-13.
31. Adema, K. N. S. (2015) *Photodegradation of Polyester-Urethane Coatings*, Technische Universiteit Eindhoven, Netherlands.
32. Allen, N. S., Chirinis-Padron, A., and Henman, T. J. (1985) The Photostabilisation of Polypropylene: A Review. *Polymer Degradation and Stability*, **13**, 31-76.
33. White, J. R., and Turnbull, A. (1994) Weathering of polymers: mechanisms of degradation and stabilization, testing strategies and modelling. *Journal of Materials Science*, **29**, 584-613.
34. *Polyurethane Handbook*, Oertel, G. (ed.), Hanser Publishers, Germany (1985).
35. Roffey, C. G. (1982) *Photopolymerization of Surface Coatings*, John Wiley & Sons, USA.
36. Rabek, J. F. (1990) *Photostabilization of Polymers: Principles and Applications*, Elsevier Applied Science, UK.
37. Kachan, A. A., Kargan, N. P., Kulik, N. V., and Boyarskii, G. Y. (2004) Two-photon heterogeneous photodegradation of an aromatic polyurethane. *Theoretical and Experimental Chemistry*, **4**, 314-317.
38. Singh, R. P., Tomer, N. S., and Bhadraiah, S. V. (2001) Photo-oxidation studies on polyurethane coating: effect of additives on yellowing of polyurethane. *Polymer Degradation and Stability*, **73**(3), 443-446.
39. Decker, C., and Bandaikha, T. (1989) *International Conference on Advances in the Stabilization and Controlled Degradation of Polymers*, Technomic Publishing, UK, p. 143.
40. *Surface Coatings: Science and Technology*, Paul, S. (ed.), 2nd edition, Wiley, UK (1995).
41. Davis, A., and Sims, D. (1983) *Weathering of Polymers*, Elsevier, Netherlands.
42. Carduner, K. R., CarterIII, R. O., Zimbo, M., Gerlock, J. L., and Bauer, D. R. (1988) End groups in acrylic copolymers. 1. Identification of end groups by carbon-13 NMR. *Macromolecules*, **21**(6), 1598-1603.
43. Larche, J. F., Bussiere, P. O., and Gardette, J. L. (2010) How to reveal latent degradation of coatings provoked by UV-light, *Polymer Degradation and Stability*, **95**(9), 1810-1817.
44. *Characterization of Highly Crosslinked Polymers*, Labana, S. S., and Dickie, R. A. (eds.), American Chemical Society, USA (1983).
45. Schmitz, P. J., Holubka, J. W., and Xu, L. F. (2000) Mechanism for environmental etch of acrylic melamine-based automotive clearcoats: Identification of degradation products. *Journal of Coatings Technology*, **72**(904), 39-45.
46. *Service Life Prediction of Organic Coatings: A Systematic Approach*, Bauer, D., and Martin, J. W. (eds.), American Chemical Society, USA

(1999).
47. Bauer, D. R., and Mielewski, D. F. (1993) The role of humidity in the photooxidation of acrylic melamine coatings. *Polymer Degradation and Stability*, **40**, 349-355.
48. Nguyen, T., Martin, J., Byrd, E., and Embree, N. (2002) Relating laboratory and outdoor exposure of coatings III. Effect of relative humidity on moisture-enhanced photolysis of acrylic-melamine coatings. *Polymer Degradation and Stability*, **77**, 1-16.
49. Shultz, A. R. (1961) Degradation of polymethyl methacrylate by ultraviolet light. *The Journal of Physical Chemistry*, **65**, 967-972.
50. Allison, J. P. (1966) Photodegradation of poly (methyl methacrylate). *Journal of Polymer Science, Part A: Polymer Chemistry*, **4**(5), 1209-1221.
51. Torikai, A. Ohno, M., and Fueki, K. (1990) Photodegradation of poly (methyl methacrylate) by monochromatic light: Quantum yield, effect of wavelengths, and light intensity. *Journal of Applied Polymer Science*, **41**, 1023-1032.
52. Chiantore, O., Trossarelli, L., and Lazzari, M. (2000) Photooxidative degradation of acrylic and methacrylic polymers, *Polymer*, **41**, 1657-1668.
53. Decker, C., and Zahouily, K. (2002) Photostabilization of polymeric materials by photoset acrylate coatings. *Radiation Physics and Chemistry*, **63**, 3-8.
54. Decker, C. Masson, F., and Schwalm, R. (2004) Weathering resistance of water based UV-cured polyurethane-acrylate coatings. *Polymer Degradation and Stability*, **83**, 309-320.
55. Nguyen, T., Martin, J., Byrd, E., and Embree, N. (2002) Relating laboratory and outdoor exposure of coatings: II. Effects of relative humidity on photodegradation and the apparent quantum yield of acrylic-melamine coatings. *Journal of Coatings Technology*, **74**, 65-80.
56. Malshe, V. C., and Waghoo, G. (2004) Weathering study of epoxy paints. *Progress in Organic Coatings*, **51**, 267-272.
57. Huang, W., Zhang, Y., Yu, Y., and Yuan, Y. (2007) Studies on UV-stable silicone-epoxy resins. *Journal of Applied Polymer Science*, **104**, 3954-3959.
58. Bellenger, V., and Verdu, J. (1983) Photooxidation of amine cross-linked epoxies I. The DGEBA-DDM system. *Journal of Applied Polymer Science*, **28**, 2599-2609.
59. Kim, H., and Urban, M. W. (2000) Molecular level chain scission mechanisms of epoxy and urethane polymeric films exposed to UV/H_2O. Multidimensional spectroscopic studies, *Langmuir*, 16, 5382-5390.
60. Rivaton, A. Moreau, L., and Gardette, J-L. (1997) Photo-oxidation of phenoxy resins at long and short wavelengths-I. Identification of the

photoproducts. *Polymer Degradation and Stability*, **58**, 321-331.
61. Mailhot, B., Morlat-Therias, S., Bussiere, P-O., and Gardette, J-L. (2005) Study of the degradation of an epoxy/amine resin, 2. *Macromolecular Chemistry and Physics*, **206**(5), 585-591.
62. Liu, F., Yin, M., Xiong, B., Zheng, F., Mao, W., Chen, Z., He, C., Zhao, X., and Fang, P. (2014) Evolution of microstructure of epoxy coating during UV degradation progress studied by slow positron annihilation spectroscopy and electrochemical impedance spectroscopy. *Electrochimica Acta*, **133**, 283-293.
63. Rabek, J. F. (1995) *Polymer Photodegradation: Mechanisms and Experimental Methods*, Springer, Netherlands.
64. Bellinger, V., Bouchard, C., Claveirolle, P. and Verdu, J., (1981) Photo-oxidation of epoxy resins cured by non-aromatic amines. *Polymer Photochemistry*, **1**(1), 69-80.
65. Bellinger, V., and Verdu, J. (1985) Oxidative Skeleton Breaking in Epoxy-Amine Networks. *Journal of Applied Polymer Science*, **30**(1), 363-374.
66. Gerlock, J. L., Smith, C. A., Nunez, E. M., Cooper, V. A., Liscombe, P., Cummings, D. R., and Dusibiber, T. G., (1996) Measurements of chemical change rates to select superior automotive clearcoats. *Advances in Chemistry*, **249**, 335-347.
67. Penon, M. G., Picken, S. J., Wűbbenhorst, M., and van Turnhout, J. (2007) Dielectric sorption analysis of pigmented epoxy coatings UV degraded at elevated pressures. *Polymer Degradation and Stability*, **92**(10), 1857-1866.
68. Santos, D., Costa, M. R., and Santos, M. T. (2007) Performance of polyester and modified polyester coil coatings exposed in different environments with high UV radiation. *Progress in Organic Coatings*, **58**(4), 296-302.
69. *ISO 4628, Paints and Varnishes: Evaluation of Degradation of Coatings. Designation of Quantity and Size of Defects, and Intensity of Uniform Changes in Appearance*, International Organization for Standardization (2016). Online: https://www.iso.org/standard/64877.html [accessed 21st March 2019].
70. Maetens, D. (2007) Weathering degradation mechanism in polyester powder coatings. *Progress in Organic Coatings*, **58**, 172-179.
71. Molhoek, L., Posthuma, C., and Gijsman, P. (2013) Weathering well. *European Coatings Journal*, 79-82.
72. Adema, K. N. S., Makki, H., Peters, E. A. J. F., Laven, J., van der Ven, L. G. J., van Benthem, R. A. T. M., and de With, G. (2014) Depth-resolved infrared microscopy and UV-VIS spectroscopy analysis of an artificially degraded polyester-urethane clearcoat. *Polymer Degradation and Stability*, **110**, 422-434.
73. Mitra, S., Ahire, A., and Mallik, B. P. (2014) Investigation of accelerated aging behaviour of high performance industrial coatings by

dynamic mechanical analysis. *Progress in Organic Coatings*, **77**(11), 1816-1825.

7

Graphene for Corrosion Protection

7.1 Introduction

Corrosion is an electrochemical process having deterioration effect on the metal or alloy. For iron, corrosion produces porous and pervious film which is composed of different forms of iron oxide. It can be seen from Figure 7.1 that redox reactions occur on the surface during this process. The sodium and chloride ions act as electrolyte, where chloride ions accelerate the corrosion process by destroying any type of passivity, thus, increasing the active corrosion rate. In this case, the accelerating corrosion process involves the dissolution of iron oxide film and, in addition, sodium and chloride ions also enhance the transportation of electrons [1,2].

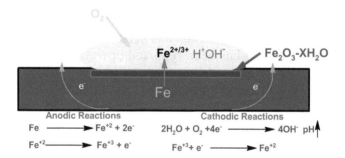

Figure 7.1 Schematic diagram of iron corrosion process; production, and consumption of electrons resulting in corrosion products.

The environmental constraints on using chromium (VI) based coatings promoted the development of non-hazardous organic and inorganic anti-corrosion pigments incorporated in polymer coatings [3]. Graphene sheets are one-atom-thick two-dimensional layers of sp^2-bonded carbon having a variety of remarkable properties and can enhance properties of polymers such as electrical and thermal

Ali U Chaudhry[a,b,*], Brajendra Mishra[b,c] and Vikas Mittal[a]
[a]Khalifa University of Science and Technology, Abu Dhabi, UAE; [b]Colorado School of Mines, USA; [c]Worcester Polytechnic Institute, USA
*Current address: Texas A&M University, Qatar
© 2021 Central West Publishing, Australia

conductivity, gas impermeability and mechanical properties, etc. [4,5]. Recently, polystyrene/graphene nanocomposites showed superior anti-corrosion properties with the incorporation of 2 wt% of filler owing to excellent barrier properties [6]. Similar results were shown for the composites of silane modified reduced graphene oxide (r-GO)/polyvinylbutyral (PVB) [7] and graphene/pernigraniline/PVB [2,8]. In the same manner, many studies have shown the single time barrier properties of stand-alone graphene films on the surface of aluminum, where excellent protection was shown after 0.5 hours of immersion in chloride environment [9]. Likewise, Raman *et al.* [5,10] measured the anti-corrosion properties of graphene film on copper after one hour of immersion. Further, the barrier properties of composites depend on many factors such as I) nanoscale level dispersion and distribution of fillers, II) interfacial compatibility of polymer and filler phases, and III) polarity match between the filler surface and the polymer chains. The full advantage of nano-fillers can only be achieved by considering above factors which could lead to uniform transfer of chemical, physical and mechanical properties of filler to the host polymer matrix [11-21]. Further, the role of conducting polymers, especially polyaniline (PANI), for corrosion protection of ferrous and non-ferrous metals has been vigorously studied [15]. For this purpose, PANI has been reported to provide corrosion protection of metals either as a neat film or as resin blended coatings [14]. Modification of inorganic pigments with PANI to achieve synergistic anti-corrosion effect has also been reported in literature [14]. Kalendova *et al.* [22] proposed a method to combine the use of inorganic pigments and PANI so as to address the problems associated with resin blended coatings, i.e., inefficient PANI distribution, lack of excellent polymer-polymer contact, poor substrate adhesion and change in volume of PANI due to redox reaction. Four pigments specularite (Fe_2O_3), goethite (FeO(OH)), talc ($Mg_3(OH)_2$-(Si_4O_{10})) and graphite were surface modified with PANI and subsequently blended with an epoxy binder. Better corrosion resistance was observed in all PANI coated pigments and PANI modified graphite exhibited excellent corrosion inhibition due to improvement in conductivity which promoted redox reactions between iron and PANI or oxygen and PANI. Sathiyanarayanan *et al.* [11,23] modified TiO_2 and Fe_2O_3 with PANI and observed enhanced corrosion protection of steel as compared to pigments without polymer modification. The results were attributed towards the formation of the passive film along with the iron-phosphate salt film on the iron surface. Brodinova *et al.* [24] also reported the presence of PANI filled the pinholes present in the coating and also formed a better interconnection between inorganic pigments and resin. Similarly, Wu *et al.* [25] reported hybrid coating of PANI-layered zinc nickel ferrites and organically modified silicate. The film was deposited using the spin coating method on aluminum alloy. The anti-corrosion performance of the hybrid film was improved due to the denser configuration of organically modified silicate due to the incorporation of nickel zinc ferrites/PANI.

The work presented here deals with the investigation of graphene as anti-corrosion filler for polymer coatings that can be used as a replacement of chromates and other hazardous pigments. Four types of studies have been selected depending on their protection mechanism generated through use of graphene in corrosive solution as well as unmodified graphene and functionalized graphene in polymer coatings. For solution properties, graphene oxide was observed to have no direct effect on the corrosion properties of steel. The effect of unmodified graphene concentration in polymer coatings was also analyzed. Graphene increased the anti-corrosion abilities of coatings when used at higher concentration, but for the shorter periods of time. Further, the effect of the modified graphene nanoplatelets was also studied, which exhibited better anti-corrosion properties of polymer coatings. To impart the electrochemical properties to the graphene, the modification was performed using polyaniline. The long time immersion exhibited that polyaniline-modified graphene had enhanced anti-corrosion properties in coatings owing to the synergistic effect [26].

7.2 Effect of Graphene on Electrochemical Properties of Carbon Steel in a Saline Media

In this work, the electrochemical properties of steel in the presence of suspended nano-graphene oxide in saline media have been studied [2]. The effect of the GO concentration (0–15 ppm) on the electrochemical properties of steel has been evaluated using electrochemical impedance spectroscopy (EIS), in addition to the morphological characterization using microscopy.

7.2.1 Materials and Methods

Nano-graphene oxide aqueous solution (concentration 1g/L, pH 2.9, purity >99%) was purchased from Graphene Supermarket, USA and used as received. The industrial steel used in this study was cut from a pipeline. API-5L X80 steel coupons were machined to 10 × 10 × 4 mm dimensions and a tap and drill hole of 3-48 tpi (threads per inch) was drilled to one long side of the coupon. Machined carbon steel was used as the working electrode and the exposed surface area was 3.4 cm². The specimens were surface finished using different grades of SiC grit papers (up to 240 grit) to ensure the same surface roughness [27,28], followed by cleaning and degreasing with industrial grade acetone and ethanol followed by drying in air. To evaluate the protection behavior of nano-GO, solution was prepared in 3.5 wt% NaCl with varying concentration of GO, i.e., 0-15 ppm.

A three-electrode cell assembly consisting of steel coupon as the working electrode (WE), graphite as the counter electrode (CE) and a saturated calomel electrode (SCE) as reference electrode (RE) were used for the electrochemical measurement. Electrochemical testing was performed in a closed system under naturally aerated conditions using a Gamry 600 potentiostat at room temperature. Impedance measurements were performed as a function of open circuit potential after five hours from the time of immersion. The frequency sweep was performed from 10^5 to 10^{-2} Hz at 10 mV AC amplitude. To simulate the electrochemical interface, EIS data was analyzed with Echem analyst using circuit model (Figure 7.2) having electrical equivalent parameters, where R_{ct} is the charge transfer resistance, L is the inductor and CPE is the constant phase element.

Accordingly, the impedance can be represented by using the following equation:

$$Z(CPE) = Y_0^{-1}[j\omega]^{-n}$$

where Y_o is the CPE constant, ω is the angular frequency (rad/s), j=√-1 and n is another CPE constant that varies from 1 to 0 for pure capacitance and pure resistor, respectively.

The double layer capacitance C_{dl} was calculated using the following equation:

$$C_{dl} = Y_0 \left[jw'' \right]^{n-1}$$

where ω" is the frequency found at the maximum of the imaginary part of the impedance, Z".

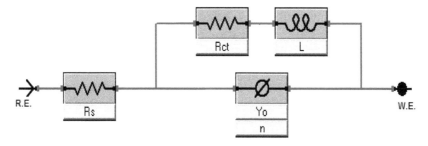

Figure 7.2 A representative circuit model used to model the electrochemical impedance spectroscopy [2].

Steel coupons were carefully disengaged from the cell assembly, dried and observed under the microscope. Field emission scanning electron microscopy (FE-SEM) using JEOL JSM-7000F was performed to evaluate surface morphology. Energy dispersive spectroscopy (EDS) was examined at 5kV under high vacuum at a working distance of 10 mm for elemental composition of the corrosion products.

7.2.2 Results and Discussion

The behavior of electrochemical processes at electrical double layer such as charge transfer resistance and ions adsorption across the electrode/electrolyte interface can be found in Table 7.1 and Figure 7.3. The figure and table depict the typical set of Nyquist plots and

Table 7.1 Electrochemical impedance spectroscopy parameters in 3.5 wt% after five hours immersion [2]

Conc. ppm	R_{ct} $(\Omega cm^2) \times 10^3$	Y_o $(\Omega^{-1}s^n)$ $\times 10^{-03}$	n	C_{dl} (Fcm^2) $\times 10^{-04}$	L (Hcm^2) $\times 10^{03}$
Blank	1.27	0.72	0.800	6.57	2.9
3	1.50	0.68	0.776	5.86	5.2
9	1.86	1.28	0.773	9.77	8.2
12	2.41	1.50	0.711	9.97	9.8
15	3.27	1.20	0.727	7.63	18.5

modeled EIS plots (Bode and Nyquist) with model circuit where charge transfer resistance was calculated from the diameter of the real part of the semicircle. It can be seen that charge transfer resistance increased with the GO concentration in the solution. It can also be observed from Figure 7.3 that profile of the Nyquist plots remained similar as the concentration of GO increased which indicated that no effect on the corrosion mechanism of carbon steel occurred with the addition of GO. Further, the double layer capacitance also increased with the GO concentration and in some cases, it was more than the blank solution, i.e., 9 and 12 ppm. Similarly, the value of n was also observed to decrease as the GO concentration increased, which was also a measure of the surface inhomogeneity; the lower is its value, the higher is the surface roughening of the metal/alloy [29]. Moreover, Y_o value also increased with the increased concentration of GO; the higher Y_o value shows that more surface area is available for the electrochemical reaction [30] due the presence of Cl⁻ ions which increases the film free area [31]. In case of blank solution, a porous corrosion product iron oxide was formed which increased the C_{dl} owing to dielectric effect as given by following equation:

$$C = \frac{\varepsilon \varepsilon_o r}{d}$$

where d is charge separation distance, ε is relative dielectric constant, ε_o is permittivity of free space, A is surface area and C is capacitance. The dissolution of the pervious layer is performed by the chloride

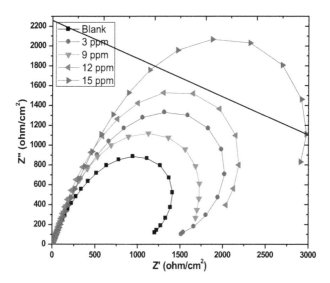

Figure 7.3 Nyquist plots recorded after 5 hour immersion in 3.5 wt% NaCl with blank and different concentrations of GO [2].

ions, however, on the addition of GO to the solution, these act as an anionic surfactant which decreases the solubility of NaCl in solution [32] and hence leads to the precipitation of salt on the working electrode. The precipitation creates porous and inhomogeneous layers which allow the availability of the corrosive solution to the working electrode. However, the precipitation appears to render more or less corrosion protection to the metal below by impeding the transportation of reactants and products among the solution and the metal [33], which results in the increment of charge transfer resistance efficiency up to 70%. Further, there was no second arc seen in the Nyquist plots which indicated that the layer forming on the surface was porous [34,35]. Nyquist plots also showed the inductance loop in the intermediate and low-frequency domain which was mainly ascribed to the occurrence of an adsorbed intermediate on the surface due to chloride ion adsorption on the electrode surface [31]. The total impedance at intermediate and low frequencies was calculated from charge transfer resistance and inductive element in series. The inductive behavior due to adsorption can be defined as L=Rτ, where τ represents the relaxation time for adsorbed species at the working electrode [36]. This can be manifested from Table 7.1 that inductance increased with increasing concentration of GO, thus, exhibiting increased absorption of salt at the working surface.

The reactions occurring at anodic and cathodic curves in 3.5 wt% NaCl are given as [37]:

$$Fe \rightarrow Fe^{2+} + e^- \, (anodic-reactions)$$

$$Fe^{2+} \rightarrow Fe^{3+} + e^-$$

$$2H^+ + 2e^- \rightarrow H_2 \, (cathodic-reaction)$$

$$2H_2O + 2e^- \rightarrow H_2 + 2OH^- \, (Neutral)$$

where the formation of a compact and thick layer of precipitates slows down the anodic and cathodic reactions. As a result, decrease in the corrosion current can be seen at higher applied potentials. Figure 7.4 shows the morphology of the outermost corrosion products layer where the surface morphology became more compact on increasing GO concentration. This phenomenon indicated that the presence of GO fostered the formation of NaCl layer on the surface which slowed down the corrosion process. The element analysis for the layers for 9 ppm, 12 ppm and 15 ppm of GO also exhibited that as the amount of added GO increased, NaCl layer became richer and iron oxide reduced.

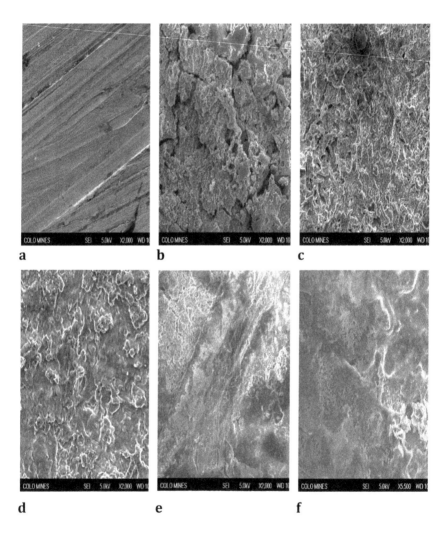

Figure 7.4 SEM micrograph of carbon steel surface after electrochemical testing in 3.5 wt% NaCl, (a) clean surface, (b) blank, different concentrations of GO (c) 3 ppm, (d) 9 ppm, (e) 12 ppm and (f) 15 ppm [2].

7.3 Effect of Unmodified Graphene Platelets (UGP) on Electrochemical Properties of Polymer Coatings

In this study, the time-dependent anti-corrosion properties of UGP based PVB composite coating on carbon steel in 0.1M NaCl aqueous solution were measured [5]. Carbon steel samples were coated with thin film of self-crosslinked composites of PVB and UGP. The corrosion properties were measured using electrochemical techniques after an immersion time of 1 and 26 h to differentiate between the corrosion barrier and corrosion promoting phenomena associated with graphene-based composites coatings.

7.3.1 Materials and Methods

Model coatings were prepared by dissolving 2000 ppm PVB (Figure 7.5; 0.2 wt% of methanol weight) and 300 ppm SDS (0.03 wt% of methanol weight) in 50 mL (39.6 g) methanol with continuous stirring for 24 h, followed by sonication in a sonicator bath. SDS was used as a dispersant and used in all coatings.

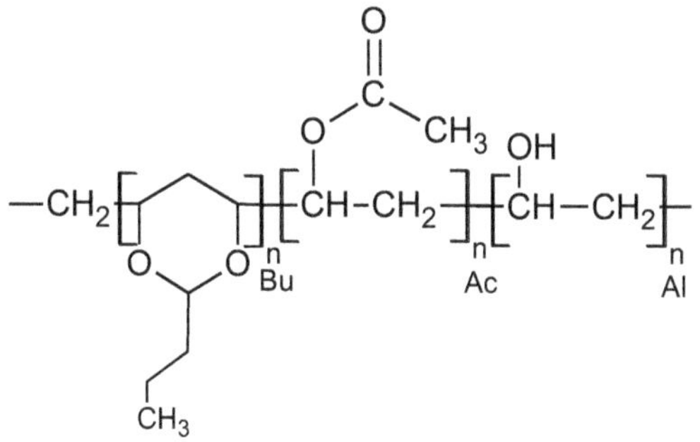

Figure 7.5 Structure of Butvar B-98 (Bu: butyral, Ac: acetate, Al: alcohol).

Similarly, two different concentrations of UGP powder, i.e., 1000 ppm (0.1 wt% of methanol weight, G-1) and 2000 ppm (0.2 wt% of methanol weight, G-2) were subjected to sonication in 50 mL (39.6 g) methanol with 300 ppm (0.03 wt% of methanol weight) SDS for 1 h. PVB was added to the UGP dispersion and shaken for 72 h to generate a uniform dispersion of UGP in the PVB solution [38]. Using dip coater, carbon steel substrates were coated with PVB-UGP dispersion by immersion and withdraw speed of 50 and 200 mm/min respectively. Samples were immersed in the solution for 1 min. Three coats were applied for each sample in a similar manner with an interval of 20 min. Further, the samples were dried at room temperature for three days followed by baking in an air circulating oven at 175 °C for 2 h to generate final coating with a thickness in the range of 70±3 µm [12].

A flat cell assembly with working volume of 250 mL (Figure 7.6), consisting of carbon steel coupon as the working electrode (WE) with exposing area 2.6 cm^2, graphite plate as counter electrode (CE) having dimensions of 25×25×5 mm with exposing area 2.6 cm^2, and a silver/silver chloride electrode as reference electrode (RE), were used for the electrochemical measurements [12]. Carbon steel panels were surface finished using different grades of SiC grit papers from 240 up to 600 grit, polished to a mirror finish followed by cleaning and degreasing with industrial grade acetone and drying in air. Before coating, specimen were treated with 2% nital for 1 min and used immediately without any further treatment [39].

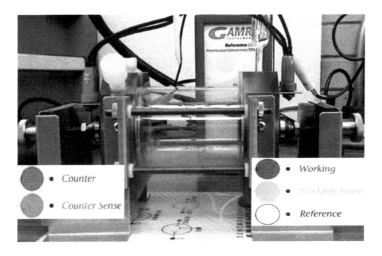

Figure 7.6 Electrochemical flat cell setup.

Electrochemical testing was performed in a closed system under naturally aerated conditions using a Gamry 600 potentiostat/galvanostat/ZRA at room temperature. Corrosion studies were carried out in 0.1 M NaCl conditions [39]. Impedance measurements were performed vs. E_{OCP} at 1 and 26 h from the time of immersion. The frequency sweep was performed from 10^5 to 10^{-2} Hz at 10 mV AC amplitude. The Bode plots were modeled with monophasic circuit model used to fit EIS data as resistor and capacitors as shown in Figure 7.7

(a-b). For the description of a frequency independent phase shift between the applied AC potential and its corresponding current response, a constant phase element (CPE) was used, where impedance of the CPE is given the equation mentioned earlier [3,39]. In the equation, n is a measure of surface inhomogeneity; the lower is its value, the higher is the surface roughening of the metal/alloy [29].

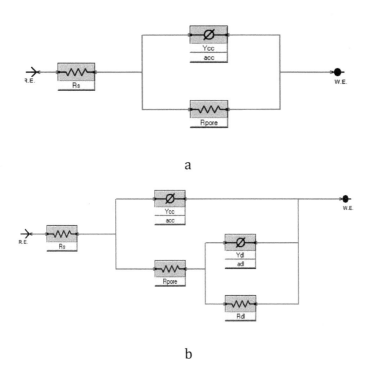

Figure 7.7 Circuit model after (1) 1 h and (b) 26 h of immersion.

At 1 and 26 h, the alloy exhibits a one-time constant impedance response for all samples. This behavior can be easily noticed in the phase angle Bode curves as a single hump for dominant one-time constant phase. To simulate the electrochemical interface, EIS data was analyzed with Echem Analyst using circuit model having electrical equivalent parameters accordingly. The capacitance C was calculated using the same equation mentioned earlier. The resistance efficiency, ER was calculated as indicated by equation [40]:

$$ER = [R'-R]/R'$$

where R' and R are the resistance value with and without UGP respectively.

Transmission electron microscopy (TEM) imaging was performed to characterize the GP. FEI Philips C200 TEM with 200 kV was used. The samples were prepared by dispersing approximately 1 mg of GP in 10 mL of acetone and sonicating for 30 min in a water bath at room temperature. One drop of the suspension was then deposited on a 400-mesh copper grid covered with a thin amorphous film to view under the microscope [2].

The bulk conductivity of the GP sheet was found to be $5.78 \times 10^{+03}$ S/m which was measured using four probe methods and had good agreement with published literature [41]. The GP sheet was obtained by pressing powder GP in a sheet form on Teflon sheet using 10 lb$_f$ for 5 min. The sheet bulk resistance was measured using four probe methods at different areas on GP sheet using 200 mV and 4.53×10^{-03} Ampere. The conductivity of sheet was calculated according to following equations:

$$\rho(\Omega.m) = R_s(sheet\ resistance, \frac{\Omega}{\blacksquare}) \times T_s(sheet\ thickness, m)$$

$$\sigma\ (conductivity, \frac{S}{m}) = \frac{1}{\rho(\Omega.m)}$$

7.3.2 Results and Discussion

The conductivity of the sheet indicated that GP had very high conductivity which could also change the conductivity of the coatings. Thus, higher amount of graphene will produce polymer coatings with higher conductivity which may affect the corrosion process happening on the surface. Figure 7.8 shows characteristic low-resolution TEM image of graphene nanoplatelets (sheets) indicating a flaky and transparent structure with wrinkles and folding on the surface [42].

The scheme in Figure 7.9 describes general physicochemical and electrochemical mechanisms of corrosion protection and promotion respectively due to the incorporation of GP in organic coatings [43]. The short-term protection is explained by a physicochemical mechanism which is generally associated with the obstruction of corrosive agents. This effect can be expected to enhance significantly by incorporating plate like reinforcements, thus, increasing the diffusion length for corrosion agents to reach the defects through microscopic pores. For reinforcement-free organic coatings, the basic mechanism is a separation of the metal surface from the environment [5].

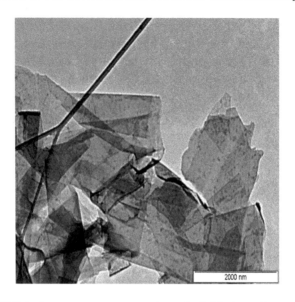

Figure 7.8 TEM of an aggregate consisting of a folded graphene nanoplatelet.

The permeability of organic coatings also depends on the nature of the binder matrix [5]. In this study, crosslinked film of PVB coating was used in order to have enhanced barrier nature of binder. The long term corrosion promotion effect was explained by the 'active' nature of UGP in the coatings. As water molecules start to accumulate at the interface, which facilitates the corrosion under the coating, graphene nanoplatelets stimulate the electrochemical reaction due to the conductivity of electrons at the interface [44]. This effect was also explained by the addition of carbon black to the zinc filled coating where carbon black was observed to promote the corrosion and acted as a perfect cathode for zinc. The addition of carbon black also improved electrical connections between the zinc particles promoting the galvanic effect [45]. In addition, it was also reported that carbon increased the porosity of the organic coating and increased the absorbance of water, thus, promoting the corrosion [46]. It was also reported that 1 g of reduced graphene (rG) sorbed 14 g of water [47].

EIS was used to examine the characteristics of defects arising in PVB and PVB/UGP composites coatings presented by complex plane plots. The coating exhibited the porous structure and the non-ideal capacitive behavior (Figure 7.9 and Table 7.2). The pore resistance of coating reduced with the addition of UGP which indicated the increased ionic resistance to current flow between the bulk and interface [48]. At the initial stage of immersion (1 h), G-2 coating exhibited quasi-ideal resistive behavior. The pore resistance of G-2 was observed to be very high as shown by half semicircle of Nyquist plot indicating the strict barrier nature of the coating, resulting in the hindrance of faradic reactions. As shown by the schematic, this barrier nature could be explained by the low diffusivity of corrosion

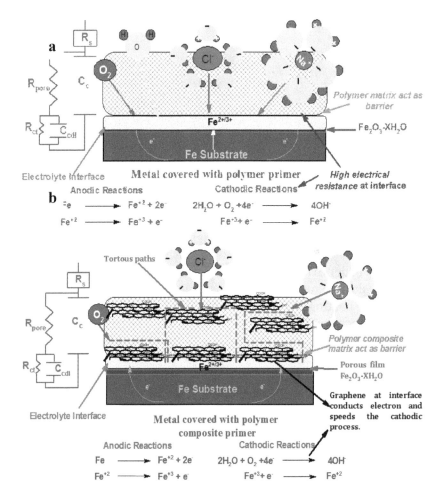

Figure 7.9 Scheme of corrosion protection phenomena in the (a) absence and (b) presence of UGP in PVB with electrochemical corrosion models for 26 h immersion in 0.1 NaCl [5].

reactants due to GP sheet-like structure (Figure 7.8). The corrosive solution has to adopt longer diffusion and tortuous paths to reach at the metal surface which in results slows down the corrosion rate. In

Table 7.2 Electrochemical parameters obtained from EIS for different samples [5]

T (h)	Sr. No.	$^b R_{pore}$ $\Omega.cm^2$ ×10³	$^c C_c$ F.cm² ×10⁻⁰⁴	$^d R_{ct}$ $\Omega.cm^2$ ×10³	$^e C_{dl}$ F.cm² ×10⁻⁴
1	PVB	1.17	4.47	-	-
	G-1	2.22	4.11	-	-
	G-2	3.98	4.34	-	-
26	PVB	0.68	4.84	1.05	1.07
	G-1	0.62	6.24	0.82	2.56
	G-2	0.65	9.36	1.04	3.63

case the defects are created by the incorporation filler due to the poor dispersion and agglomerations of fillers, this phenomena may cause inverse results.

The pore resistance of coatings was calculated by using the following equation [10]:

$$|Z| = \sqrt{Z_{real}^2 + Z_{imaginary}^2}$$

at lowest frequency, where $Z_{imaginary} \to 0$ giving $|Z| = Z_{real}$ in Bode plot. It is also evident from Bode plots of Figure 7.10 a that the pore resistance of G-1 and G-2 was at least 50% and 70% respectively greater than that of PVB coated steel after 1 h of immersion. With increasing exposure time, i.e., 26 h of immersion, it was taken into account that corrosive media had accumulated on the carbon steel surface through the coating. EIS data from Figure 7.10 a and Table 7.2 shows the variations in the impedance model as a response to the intact area. The change in the EIS plots indicated that the model used for 1 h of immersion did not satisfy for longer period immersion. It can be noticed that coating capacitance of G-1 and G-2 was increased by 23% and 50% respectively. Similarly, interface capacitance of G-1 and G-2 had an increment of 58% and 71% respectively, showing enhancement of corrosion. Although, there was not much difference in

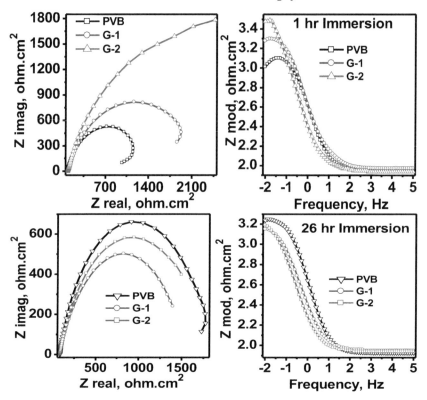

Figure 7.10 EIS magnitude spectra of coated carbon steel after a) 1 hour and b) 26 hour immersion in 0.1 M NaCl [5].

the coating and charge transfer resistance, but these parameters showed decreased value for G-1 and G-2. The presence of water in the pores can change the dielectric constant of the coating or interface and can be determined by the capacitance [48]. The increased capacitance values with increased exposure time for G-1 and G-2 indicated that presence of corrosive reactant and products in coatings and coating/metal interface changed the local effective dielectric constant and water uptake properties according to following equations [13]:

$$C = \frac{\varepsilon \varepsilon_0 A}{d} \text{ and } \theta \text{ (percent)} = \frac{\log (C_t/C_o)}{\log (\varepsilon_w)}$$

where d is the charge separation distance, ε is the relative dielectric constant, ε_o is the permittivity of free space, A is the surface area, C is the capacitance, θ is the percentage water uptake, C_t is the coating capacitance at 26 h, C is the initial coating capacitance at 1 h, and ε_w is the relative permittivity of water taken as 80.1 at 20 °C, where relative permittivity is a dimensionless value. It can see from Figure 7.11 that addition of UGP enhanced the water uptake percentage to 9.5% for PVB to UGP ratio of 2:1 and almost double with the 1:1 ratio of PVB to UGP. These results had conformity with the previous work reported on coatings containing carbon black [46]. SEM images were taken at the cross-sectional area to study the morphological alterations of coatings at metal/coating interface. The corrosion effects can be confirmed in Figure 7.11 where larger defects could be observed for G-1 and G-2. The approximate defect size of coating for PVB and GP was around 5 μm and 10 μm, respectively. The larger defects in the GP containing coatings indicated that water uptake properties of coatings increased, which in turn produced accelerated coating deterioration.

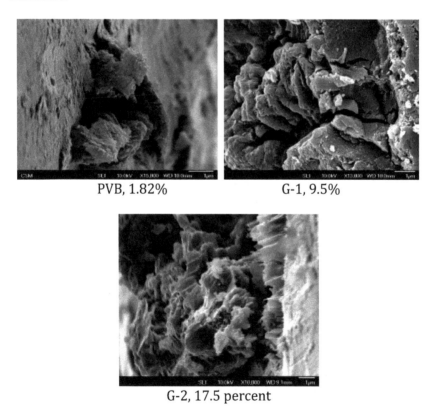

Figure 7.11 Cross-sectional area of coating showing coating/metal interface and water uptake θ (percent) after immersion of 26 h [5].

7.4 Effect of Functionalized Graphene Platelets on Electrochemical Properties of Polymer Coatings

This study deals with the comparison of barrier properties of nanocomposites prepared from Haydale's graphene and thermally reduced graphene oxide [49]. In specific, four kinds of graphene, i.e., U-GP and Haydale's processed graphene nanoplatelets produced using three types of process gasses, i.e., fluorine (F-GP), oxygen (O-GP), and argon (A-GP) were used.

7.4.1 Materials and Methods

PVB with trade name Butvar B-98 (molecular weight 40,000-70,000 g/mol) was purchased from Sigma-Aldrich. PVB had 18-20% hydroxyl content and 80% butyral content. M-GP (commercial name HDPlas™), with planer size 0.3-5 μm and platelets thickness < 50nm, was used as purchased from Graphene Supermarket, USA. U-GP (A-12, graphene supermarket, USA), with thickness < 3nm and planer size 2-8 μm, was used for comparison. GP amount was kept constant at 3 wt% of PVB resin (5g resin/50 ml methanol) for all the composites [12].

EIS was performed on coated carbon steel panels at different times of immersion, i.e., 1 and 12 h in 4 wt% percent NaCl. Model coatings were prepared by bath sonication of GP for 4 h in 50 ml methanol followed by 48 h of mixing with PVB in a closely tight glass flask. Carbon steel samples were polished starting from 60 to 600 grit sand paper. Using dip coater, carbon steel substrates were coated with PVB-GP dispersion by immersion and withdraw speed of 50 and 200 mm/min respectively. Samples were immersed in the solution for 1 min. Two coats were applied to each sample in a similar manner with an interval of 15 min. Further, the samples were dried at room temperature for 3 d followed by baking in an air circulating oven at 175 °C for 2 h to generate final coating with a thickness in the range of 270±3 μm. Four carbon steel samples were prepared for each PVB and composite coatings. A flat cell assembly (Figure 7.6) was used for corrosion measurements. The open circuit potential of steel samples was recorded against Ag/AgCl electrode as a reference electrode for 1 and 12 h. After the completion of each step, EIS was measured. Impedance measurements were performed vs. (E_{OCP}) after 1 and 12 h from the time of immersion. The frequency sweep was performed from 10^5 to 10^{-2} Hz at 10 mV AC amplitude.

7.4.2 Results and Discussion

TEM (Figure 7.12) of M-GP indicated different morphological features from that of UGP. Figure 7.13 (a-b) and Table 7.3 depict the phase angle and extracted model circuit parameters respectively for coated carbon steel in 4 wt % NaCl aqueous solution for different immersion time.

Phase angle exhibited distinct behavior at a various range of frequency for PVB and composites coatings. The range of frequency corresponds to the various regions in the coating, i.e., starting from coating surface to substrate surface. After 1 h of immersion, the PVB coating at higher frequency ~100 kHz exhibited a mixed capacitive and resistive behavior (~-43°), which indicated the barrier and diffusive

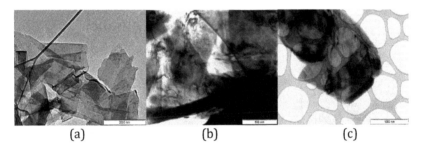

(a)   (b)   (c)

Figure 7.12 TEM images of graphene nanoplatelets; (a) U-GP, (b) A-GP and (c) F-GP [49].

nature of the coating. The same behavior continued over intermediate frequency range with a variation of theta around ~-45°±-5°, afterwards it changed to resistive behavior in the low-frequency spectrum. At frequency ~25 kHz, the slight increase in the angle ~-37° indicated dominant resistive behavior. A little bump could be observed at 12.4 Hz corresponding to the phase angle ~-46° between two minima of theta ~-49° at 100 Hz and ~-48.8° at 2.5 Hz. The continuous resistive and capacitive behavior ranging from higher to intermediate frequency could be an indication of constrained diffusion of electrolyte through the coating corresponding to porous bounded Warburg (PB-W) as mentioned in Gamry Echem Analyst™ manual. The major transition from mixed resistive-capacitive behavior to resistive behavior occurs at ~0.9 Hz corresponding to the metal/interface and metal characteristics. Some literature reports (especially Zhang et al.) used PB-W element to model the diffusion limited behavior in fuel cell containing meshes of steel [50-52]. The addition of

U-GP changed the phase angle behavior considerably. At frequency ~100 kHz, comparatively lower phase angle ~-68° indicated high capacitive nature of coating followed by a dominant resistive behavior owing to higher phase angle ~-30° at 800 Hz. The behavior changed to dominant capacitive behavior ~-57° at ~12.5 Hz. This trend indicated a large amount of trapped electrolyte in the coating which was further unable to respond to the AC frequency at the intermediate frequency. This behavior could be modeled using Gerischer element [53] used for modeling a porous electrode as mentioned in Gamry Echem Analyst™ manual. Table 7.3 shows that addition of U-GP did not significantly improve the pore resistance and admittance of coating. The admittance value was very high as compared to PVB coating.

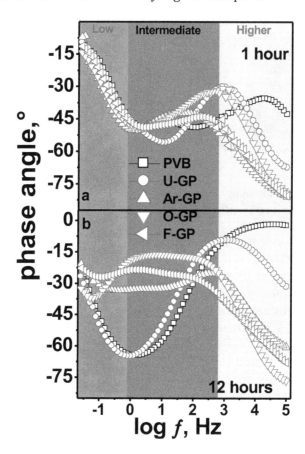

Figure 7.13 Phase angle behavior for PVB and GP-PVB nanocomposites coating on carbon steel recorded after 1 h and 12 h of immersion in 4 wt% NaCl [49].

Graphene for Corrosion Protection

Table 7.3 EIS parameters extracted using different circuit models

Model Circuits	EIS parameters	PVB	U-GP-PVB	O-GP-PVB	Ar-GP-PVB	F-GP-PVB
	PVB 1h	$R_{pore} \times 10^{03}$	2.93	2.75	26.9	87.4	49.0
	M-GP 1h	$R_{ct} \times 10^{04}$	2.67	21.6	618	456	326
		$Y_{admit.} \times 10^{-05}$	0.32^W	8.52^G	0.18^G	0.09^G	0.13^G
	PVB 12h	$R_{pore} \times 10^{04}$	1.01	1.28	1.36	1.52	4.40
	M-GP 12h	$R_{ct} \times 10^{03}$	4.51	12.6	10.4	24.6	34.1
		$Y_{admit.} \times 10^{-05}$	-	79.6	2.87	2.22	9.22

*Resistance Ohm.cm^2, Y(admit.) S*s$^{(1/2)}$/cm^2*

However, there was a significant improvement in the R_{ct} indicating that corrosive electrolyte was although already in the coating, but unable to reach the interface. Upon adding graphene produced from Haydale's process to PVB, the composite coating showed remarkable enhancement in anti-corrosion properties owing to improved diffusion behavior indicating improved dispersion and interaction with the PVB resin.

In the case of nanocomposites produced from F-GP/PVB, at ~100 kHz frequency, the coating exhibited a highly capacitive behavior indicated by low phase angle, i.e., ~-81° which indicated improved barrier properties. The behavior gradually changed to capacitive-resistive behavior at ~400 Hz frequency. Afterwards, frequency independent behavior could be seen over a range of frequency from 400 Hz to 1.6 Hz with constant phase angle ~-45°±3° indicating a diffusion limited process, a trend shown by all of other cases. This kind of impedance behavior enhanced the RC by 28, 21 and 15 times for O-GP, A-GP, and F-GP respectively, as compared to U-GP after 1 h of immersion. After 12 h of immersion, PVB coating exhibited an entirely different behavior. At the higher frequency, it depicted resistive behavior indicating loss of barrier properties. On the other hand, at intermediate and low-frequency range, dominant capacitive hump could be seen indicating the appearance of larger blisters in the coatings [54]. Similar was the case for U-GP at intermediate and lower frequency, but it exhibited lesser capacitive behavior at a higher frequency as compared to 1 h indicating reduced barrier and adhesion. The M-GP nanocomposites still exhibited very high capacitive behavior at higher frequency indicating that the coating still retained barrier properties and was in contact with the substrate. The capacitive behavior gradually changed to more resistive (~-33° for A-GP, ~-23°for F-GP and ~-17°for O-GP) as compared to 1 h immersion and extended over a broad range of low frequencies. For O-GP, a little capacitive hump could be observed at very low frequency indicating appearance of tiny blisters. Further, the continuous resistive behavior also indicated that delaminated area was small where the corrosion was taking place. Table 7.3 also indicated that compared to U-GP, M-GP nanocomposites had very low admittance value determined from Gerischer element after 12 h.

7.5 Effect of Functionalized Graphene Platelets with Polyaniline on Electrochemical Properties of Polymer Coatings

In the current study, PANI and the nano-hybrids based on PANI-thermally reduced graphene (r-GO) were used as the anti-corrosion pigments incorporated in poly(vinyl butyral) (PVB) resin [13]. The PANI/r-GO hybrids were generated with varying r-GO/aniline ratios during the hybrid generation, so as to relate the effect of graphene content to the coating performance. The coatings were applied on carbon steel panels and the corrosion performance of these pigments was investigated in 4% w/v NaCl solution.

7.5.1 Materials and Methods

PVB (M_W 70,000-100,000) resin with broad molecular weight distribution, containing ~20% vinyl alcohol and ~2% vinyl acetate, was obtained as granules from Polysciences, Inc. (USA). Carbon steel coupons (elemental composition: C 0.07 wt%, Mn 1.36 wt%, Ti 0.008 wt%, S 0.003 wt%, P 0.004 wt% and rest Fe) were purchased locally and were machined to 30 x 16 x 5 mm dimensions. Polyaniline modified r-GO was prepared using *in-situ* polymerization. To generate PANI/r-GO composites with different weight ratios, 0, 10, 20, 30, 40 and 50 mg of r-GO nanoparticles were added to a mixture of 1 ml of aniline and 90 ml of 1N HCl in a reaction vessel (Figure 7.14) [11]. The mixture was stirred in ice water bath (5 °C) for 1 h to obtain a uniform dispersion of r-GO. To the mixture, 100 ml pre-cooled 1N HCl solution containing 2.5 g ammonium persulfate (APS) was added dropwise. The reaction was allowed to proceed in the ice bath for 4 h. The resulting product was washed with distilled water several times and at last washed with methanol in order to eliminate oligomers and other impurities [11]. Subsequently, the sample was dried in a vacuum oven for 12 h. Based on the amount of r-GO, these hybrids were named as PANI/10r-GO, PANI/20r-GO, PANI/30r-GO, PANI/40r-GO and PANI/50r-GO.

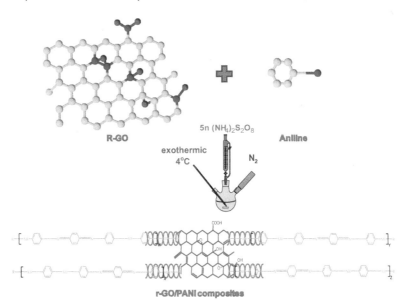

Figure 7.14 Schematic diagram of generation of PANI/r-GO composite particles.

PVB was dissolved in methanol (20 ml) under stirring for 6 h, followed by sonication in a bath sonicator for 2 h. PANI/r-GO nanoparticles of different compositions were finely powdered with a small quantity of methanol to obtain a uniform slurry. The slurry was added and mixed in the PVB solution and sonicated for 24 h to achieve uniform dispersion of nanoparticles in the PVB solution [38]. The amount of PANI/r-GO particles was fixed to 10 wt % of the amount of dry PVB. For coating of the generated PANI/r-GO/PVB formulations, carbon steel coupons with rounded corners and edges were polished by emery paper, washed with acetone, dried and weighed with an accuracy of ±1 mg. The substrates were brush coated with the PANI/r-GO/PVB formulations and dried at room temperature for 30 min. Subsequently, the coating was baked in an air circulating oven at 60 °C for 6 h and coatings with a thickness in the range of 15-20 µm were obtained.

Transmission electron microscopy (TEM) of hybrid samples was performed using EM 912 Omega (Zeiss, Oberkochen BRD) electron microscope at 120 kV and 200 kV accelerating voltage. Thin flakes of PANI/r-GO platelets were supported on 100 mesh grids sputter coated with a 3 nm thick carbon layer.

The anti-corrosion properties of the coatings were studied by immersion test (NaCl 4% w/v) according to ASTM G31. The coated coupons were placed in a specially designed set up for 800 h at 25 °C. The function of this set-up was to maintain temperature, air flow, and water level so as to attain similar conditions throughout the experiment [14]. After completing the immersion period, the samples were taken out of salt solution, cleaned and weighed again. Cleaning of the samples was achieved by removing the coating using acetone as a solvent and subsequently removing the corrosion products by a solution of 3.5 g of hexamethylenetetramine in 500 ml of hydrochloric acid diluted to 1000 ml with distilled water. The samples were dipped for 10 min in the amine solution at room temperature. Weight loss method was used to evaluate corrosion rate (CR), using the following equation:

$$Corrosion\,rate, CR = \frac{87.6\,W}{DAT}$$

where CR is corrosion rate in mm/year, W is weight loss in mg, D is the substrate density, A is the exposed surface area and T is exposure time in the salt solution in hours. The area of the exposed surface of the coated substrate was 9.60 cm^2, whereas the density of substrate was taken as 7.85 g/cm^3. The corrosion protection efficiency (CE) of anti-corrosion pigment was calculated using following equation:

$$CE = \left[\frac{CR - CR'}{CR}\right]$$

where CR and CR' are the corrosion rates without and with the anti-corrosion pigment respectively.

7.5.2 Results and Discussion

In the current study, PANI/r-GO hybrids with varying amount of graphene were incorporated in PVB matrix so as to achieve anti-corrosion coatings where both PANI and graphene synergistically contributed to the coating's performance. The morphology of the graphene platelets altered completely with PANI intercalation in the interlayers. PANI was observed to uniformly cover the surface of the platelets, which was also the reason for the absence of any graphene diffraction signal in the X-ray diffractograms. Even increasing the amount of graphene in the hybrid exhibited similar morphology. PANI chains are expected to physically adsorb on the surface of graphene nanoparticles due to interactions of N atoms with functional groups on graphene surface as well as due to van der Waals forces between PANI chains and graphene surface. This is, thus, also expected to affect the polymer crystallinity as the PANI chains would adsorb on the surface of platelets simultaneously during their synthesis, which will hinder their effective crystallization [11].

From the corrosion rate analysis, it was observed that under the test conditions used, pure PANI pigment in PVB did not provide any superior corrosion protection than pure PVB coating itself, though both improved the protection of bare metal substrate. The corrosion protection efficiency (CE) was observed to increase with increasing amount of r-GO in the coatings. These results can be attributed to the additional protection provided by r-GO, along with the intrinsic protection provided by PANI. For instance, coating with PANI/40r-GO was nearly 52% more effective than the PVB coating without pigment. Further increase in graphene content in PANI/50r-GO exhibited a marginal increase in protection efficiency to 58%, which indicated that the CE may have reached a plateau.

To further confirm these findings, optical microscopy was employed to gain insights into the change in the surface topography of the substrates after the corrosion test. Figure 7.15 shows the surface of the coatings after the corrosion test as well as removal of the coating. Substrates coated with pure PVB (Figure 7.15 (a)) and pure PANI in PVB (Figure 7.15 (b)) formulations exhibited extensive surface corrosion. It indicated that the corrosion media could penetrate faster through the coatings, thus, initiating the corrosion process. The substrate coated with PANI/10r-GO (Figure 7.15 (c)) in PVB exhibited improved anti-corrosion performance as the extent of surface corrosion was markedly reduced in comparison with Figure 7.15 (a) and (b). A further increase in the graphene concentration enhanced coatings performance further. PANI/20r-GO, PANI/30r-GO, and PANI/40r-GO coated substrates optically exhibited a high level of stability to the immersion conditions. Interestingly, the conductivity of the coatings decreased as the amount of reduced graphene was increased. It has been reported earlier that the lower conductivity of

PANI/graphene composites results probably due to decrease in the degree of doping in PANI and change in the morphology of the composites [55, 56]. Similarly, Sun et al. [8] have also reported graphene/pernigraniline composite (GPCs) coatings in PVB with reduced conductivity for the corrosion protection of copper. The GPCs were generated by in-situ polymerization-reduction/doping process and the authors concluded that pernigraniline modified graphene in PVB was able to provide effective corrosion protection.

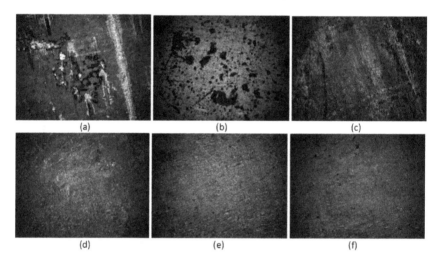

Figure 7.15 Surface of the substrates after corrosion testing and coating removal; (a) blank (pure PVB), (b) PANI, (c) PANI/10r-GO, (d) PANI/20r-GO, (e) PANI/30r-GO and (f) PANI/40r-GO. The width of the images equals 200 μm.

The corrosion protection by the coating is due to the generation of disconnection between bare metal and corrosive environment, thus, impeding the transportation of corrosive material. The protection behavior of PANI at the interface of metal/electrolyte has been described earlier in the literature [57]. In this phenomenon, the formation of metal oxide layer occurs with the aid of more noble emeraldine salt (ES) of PANI. This form of PANI further leads to lower energy state and reduces to different forms such as non-conducting leucoemeraldine base (LEB) and emeraldine base (EB). This cycle is continuous and LE form turns to ES via EB form by oxygen. This cycle can be continuous only if the barrier properties of applied coating are able to avoid the removal of H^+ (maintained acidic pH) by the surroundings. This process slows the formation of the rust at the iron surface by providing passive layers of different forms of PANI. However, as observed in the current study, pure PANI present in the PVB matrix was unable to significantly slow down the corrosion rate probably due to the lower extent of PANI-metal interface formation. In addition, the process of corrosion protection could be specifically enhanced by the addition of r-GO, as it resulted in the tortuous path for the corrosive media, thus, delaying the corrosion process. This was also confirmed from the improving corrosion protection efficiency as the amount of graphene was enhanced in the coatings. Such barrier effect of graphene-based reinforcement is called a 'passive' role. Further, the 'active' behavior of such material was realized in the coatings by the formation of 'Schottky barrier' at the interface, thus, leading to the depletion of electrons, which slowed down the corrosion half-cell reactions. Composites of PANI acted like hetro-junction, where PANI behaves as p-type while r-GO being n-type, thus, hindering the anodic and cathodic reactions respectively. The decreased conductivity of composites also impedes the transport of electron from graphene to PANI [38].

Furthermore, due to the physical adsorption of PANI on the graphene surface, the effect of graphene incorporation was synergistically enhanced by the presence of PANI due to its better interface with the metal. It has also to be noted that with these coatings systems, the corrosion reaction was not completely eliminated, however, it was significantly slowed down, thus underlining the need for the functional coatings to achieve superior material performance [6,58].

7.6 Conclusions

Graphene represents a very useful nanomaterial for the development of anti-corrosion coatings. The studies presented in this chapter confirm the high potential of graphene and its derivatives in inhibiting and delaying corrosion on metal surfaces in various corrosion environments. In this category, the specific applications or performance enhancements can also be generated by suitably modifying the graphene surface as well as optimizing its interaction with the polymers, along with nano-scale dispersion. In addition, graphene significantly impacts both short-term and ling-term anti-corrosion behavior of the coatings. Further efforts are needed to be focused on overcoming the challenges associated with large scale production of graphene as well as uniform dispersion of graphene in the coating matrices.

References

1. Jones, D. A. (2013) *Principles and Prevention of Corrosion,* Pearson Education Limited, USA.
2. Chaudhry, A. U., Mittal, V., and Mishra, B. (2015) Effect of graphene oxide nanoplatelets on electrochemical properties of steel substrate in saline media. *Materials Chemistry and Physics*, **163**, 130-137.
3. Chaudhry, A. U., Bhola, R., Mittal, V., and Mishra, B. (2014) $Ni_{0.5}Zn_{0.5}Fe_2O_4$ as a potential corrosion inhibitor for API 5L X80 steel in acidic environment. *International Journal of Electrochemical Science*, **9**, 4478-4492.
4. Kim, J., Cote, L. J., Kim, F., Yuan, W., Shull, K. R., and Huang, J. (2010) Graphene oxide sheets at interfaces. *Journal of the American Chemical Society*, **132**(23), 8180-8186.
5. Chaudhry, A.U., Mittal, V., and Mishra, B. (2015) Inhibition and promotion of electrochemical reactions by graphene in organic coatings. *RSC Advances*, **5**(98), 80365-80368.
6. Yu, Y.-H., Lin, Y.-Y., Lin, C.-H., Chan, C.-C., and Huang, Y.-C. (2014) High-performance polystyrene/graphene-based nanocomposites with excellent anti-corrosion properties. *Polymer Chemistry*, **5**(2), 535-550.
7. Sun, W., Wang, L., Wu, T., Wang, M., Yang, Z., Pan, Y., and Liu, G. (2015) Inhibiting the corrosion-promotion activity of graphene. *Chemistry of Materials*, **27**(7), 2367-2373.
8. Sun, W., Wang, L., Wu, T., Pan, Y., and Liu, G. (2014) Synthesis of low-electrical-conductivity graphene/pernigraniline composites and their application in corrosion protection. *Carbon*, **79**, 605-614.
9. Liu, J., Hua, L., Li, S., and Yu, M. (2015) Graphene dip coatings: An effective anticorrosion barrier on aluminum. *Applied Surface Science*, **327**, 241-245.
10. Singh Raman, R. K., Chakraborty Banerjee, P., Lobo, D. E., Gullapalli, H., Sumandasa, M., Kumar, A., Choudhary, L., Tkacz, R., Ajayan, P. M., and Majumder, M. (2012) Protecting copper from electrochemical degradation by graphene coating. *Carbon*, **50**(11), 4040-4045.
11. Mittal, V., Chaudhry, A. U., and Matsko, N. B. (2016) Organic functionalization of thermally reduced graphene oxide nanoplatelets by adsorption: structural and morphological characterization. *Philosophical Magazine*, **96**(20), 2143-2160.
12. Mishra, B., Chaudhry, A., and Mittal, V. (2016) Development of polymer-based composite coatings for the gas exploration industry: Polyoxometalate doped conducting polymer based self-healing pigment for polymer coatings, *Materials Science Forum*, **879**, 60-65
13. Chaudhry, A. U., and Mittal, V. (2017) Polyaniline-graphene composite nanoparticle pigments for anti-corrosion coatings. In: Conducting Polymer Composites, Mittal, V. (ed.), CWP Australia, pp. 69-94.

14. Khan, M. I., Chaudhry, A. U., Hashim, S. and Iqbal, M. Z. (2010) Investigation of corrosion-protective performance of polyaniline covered inorganic pigments. *Nucleus*, **47**(4), 287-293.
15. Khan, M. I., Chaudhry, A. U., Hashim, S., Zahoor, M. K., and Iqbal, M. Z. (2010) Recent developments in intrinsically conductive polymer coatings for corrosion protection. *Chemical Engineering Research Bulletin*, **14**(2), 73-86.
16. Luckachan, G. E., and Mittal, V. (2015) Anti-corrosion behavior of layer by layer coatings of cross-linked chitosan and poly(vinyl butyral) on carbon steel. *Cellulose*, **22**, 3275-3290.
17. Mittal, V., and Chaudhry, A. U. (2015) Polymer-graphene nanocomposites: effect of polymer matrix and filler amount on properties. *Macromolecular Materials and Engineering*, **300**(5), 510-521.
18. Mittal, V., and Chaudhry, A. U. (2015) Effect of amphiphilic compatibilizers on the filler dispersion and properties of polyethylene-thermally reduced graphene nanocomposites. *Journal of Applied Polymer Science*, **132**(35), DOI: 10.1002/app.42484.
19. Mittal, V., Chaudhry, A. U., and Khan, M. I. (2011) Comparison of anti-corrosion performance of polyaniline modified ferrites. *Journal of Dispersion Science and Technology*, **33**(10), 1452-1457.
20. Mittal, V., Chaudhry, A. U., and Luckachan, G. E. (2014) Biopolymer-thermally reduced graphene nanocomposites: Structural characterization and properties. *Materials Chemistry and Physics*, **147**(1-2), 319-332.
21. Mittal, V., Chaudhry, A. U., and Matsko, N. B. (2014) "True" biocomposites with biopolyesters and date seed powder: Mechanical, thermal, and degradation properties. *Journal of Applied Polymer Science*, **131**(19), DOI: 10.1002/app.40816.
22. Kalendova, A., Sapurina, I., Stejskal, J., and Vesely, D. (2008) Anticorrosion properties of polyaniline-coated pigments in organic coatings. *Corrosion Science*, **50**(12), 3549-3560.
23. Sathiyanarayanan, S., Azim, S. S., and Venkatachari, G. (2007) A new corrosion protection coating with polyaniline-TiO_2 composite for steel. *Electrochimica Acta*, **52**(5), 2068-2074.
24. Brodinova, J., Stejskal, J., and Kalendova, A. (2007) Investigation of ferrites properties with polyaniline layer in anticorrosive coatings. *Journal of Physics and Chemistry of Solids*, **68**(5-6), 1091-1095.
25. Wu, K. H., Chao, C. M., Liu, C. H., and Chang, T. C. (2007) Characterization and corrosion resistance of organically modified silicate–NiZn ferrite/polyaniline hybrid coatings on aluminum alloys. *Corrosion Science*, **49**(7), 3001-3014.
26. Usman, C. A. (2016) *Anti-corrosion Behaviour of Barrier, Electrochemical and Self-healing Fillers in Polymer Coatings for Carbon Steel in a Saline Environment*, Ph.D. Thesis, Colorado School of Mines, USA.

27. New Trends in Electrochemical Impedance Spectroscopy (EIS) and Electrochemical Noise Analysis (ENA), Mansfeld, F., Huet, F., and Mattos, O. (eds.), The Electrochemical Society, USA.
28. Asma, R. N., Yuli, P., and Mokhtar, C. (2011) Study on the effect of surface finish on corrosion of carbon steel in CO_2 environment. *Journal of Applied Sciences*, **11**(11), 2053-2057.
29. Chongdar, S., Gunasekaran, G., and Kumar, P. (2005) Corrosion inhibition of mild steel by aerobic biofilm. *Electrochimica Acta*, **50**(24), 4655-4665.
30. Mora-Mendoza, J. L., and Turgoose, S. (2002) Fe_3C influence on the corrosion rate of mild steel in aqueous CO_2 systems under turbulent flow conditions. *Corrosion Science*, **44**(6), 1223-1246.
31. Fekry, A. (2011) Electrochemical corrosion behavior of magnesium alloys in biological solutions. In: *Magnesium Alloys - Corrosion and Surface Treatments*, Czerwinski, F. (ed.), Intech, Croatia, doi: 10.5772/13027.
32. Zhou, X., and Hao, J. (2011) Solubility of NaBr, NaCl, and KBr in surfactant aqueous solutions. *Journal of Chemical & Engineering Data*, **56**(4), 951-955.
33. Lopez, D. A., Simison, S. N., and de Sanchez, S. R. (2003) The influence of steel microstructure on CO_2 corrosion. EIS studies on the inhibition efficiency of benzimidazole. *Electrochimica Acta*, **48**(7), 845-854.
34. Jiang, X., and Nesic. S. (2008) Electrochemical Investigation of the Role of Cl-on Localized CO_2 Corrosion of Mild Steel. *17th International Corrosion Congress, USA*. Online: http://www.corrosioncenter.ohiou.edu/documents/publications/8209.pdf [assessed 23rd April 2017].
35. Hernandez, J., Munoz, A., and Genesca, J. (2012) Formation of iron-carbonate scale-layer and corrosion mechanism of API X70 pipeline steel in carbon dioxide-saturated 3% sodium chloride. *Afinidad*, **69**(560), 251-258.
36. Fekry, A., Ghoneim, A., and Ameer, M. (2014) Electrochemical impedance spectroscopy of chitosan coated magnesium alloys in a synthetic sweat medium. *Surface and Coatings Technology*, **238**, 126-132.
37. Sarkar, P. P., Kumar, P., Manna, M. K., and Chakraborti, P. C. (2005) Microstructural influence on the electrochemical corrosion behaviour of dual-phase steels in 3.5% NaCl solution. *Materials Letters*, **59**(19-20), 2488-2491.
38. Radhakrishnan, S., Siju, C. R., Mahanta, D., Patil, S., and Madras, G. (2009) Conducting polyaniline-nano-TiO_2 composites for smart corrosion resistant coatings. *Electrochimica Acta*, **54**(4), 1249-1254.
39. Chaudhry, A. U., Mittal, V., and Mishra, B. (2015) Nano nickel ferrite

(NiFe$_2$O$_4$) as anti-corrosion pigment for API 5L X-80 steel: An electrochemical study in acidic and saline media. *Dyes and Pigments*, **118**, 18-26.
40. Sathiyanarayanan, S., Jeyaprabha, C., Muradidharan, S., and Venkatachari, G. (2006) Inhibition of iron corrosion in 0.5 M sulphuric acid by metal cations. *Applied Surface Science*, **252**(23), 8107-8112.
41. Kundhikanjana, W., Lai, K., Wang, H., Dai, H., Kelly, M. A., and Shen, Z.-x. (2009) Hierarchy of electronic properties of chemically derived and pristine graphene probed by microwave imaging. *Nano Letters*, **9**(11), 3762-3765.
42. Shao, Y., Zhang, S., Wang, C., Nie, Z., Liu, J., Wang, Y., and Lin, Y. (2010) Highly durable graphene nanoplatelets supported Pt nanocatalysts for oxygen reduction. *Journal of Power Sources*, **195**(15), 4600-4605.
43. Funke, W. (1986) How organic coating systems protect against corrosion. In: *Polymeric Materials for Corrosion Control*, American Chemical Society, USA, pp. 222-228.
44. Schriver, M., Regan, W., Gannett, W. J., Zaniewski, A. M., Crommie, M. F., and Zettl, A. (2013) Graphene as a long-term metal oxidation barrier: Worse than nothing. *ACS Nano*, **7**(7), 5763-5768.
45. Marchebois, H., Touzain, S., Joiret, S., Bernard, J., and Savall, C. (2002) Zinc-rich powder coatings corrosion in sea water: influence of conductive pigments. *Progress in Organic Coatings*, **45**(4), 415-421.
46. Nazeri, M. F. M., Suan, M. S. M., Masri, M. N., Alias, N., and Mohamed, A. A. (2012) Corrosion studies of conductive paint coating using battery cathode waste material in sodium chloride solution. *International Journal of Electrochemical Science*, **7**, 6976-6987.
47. Iqbal, M., and Abdala, A. (2013) Oil spill cleanup using graphene. *Environmental Science and Pollution Research*, **20**(5), 3271-3279.
48. Macdonald, D. D., and McKubre M. C. H. (1981) Electrochemical impedance techniques in corrosion science, ASTM, USA. Online: https://www.astm.org/DIGITAL LIBRARY/STP/PAGES/STP28030S.htm [assessed 23rd April 2017].
49. Chaudhry, A. U., Mittal, V., and Mioshra, B. (2017) Impedance response of nanocomposite coatings comprising of polyvinyl butyral and Haydale's plasma processed graphene. *Progress in Organic Coatings*, **110**, 97-103.
50. Zhang, F., Merrill, M. D., Tokash, J. C., Saito, T., Cheng, S., Hickner, M. A., and Logan, B. E. (2011) Mesh optimization for microbial fuel cell cathodes constructed around stainless steel mesh current collectors. *Journal of Power Sources*, **196**(3), 1097-1102.
51. Lamaka, S. V., Zheludkevich, M. L., Yasakau, K. A., Montemor, M. F., and Ferreira, M. G. S. (2007) High effective organic corrosion inhibitors for 2024 aluminium alloy. *Electrochimica Acta*, **52**(25), 7231-

7247.
52. Kuzum, D., Takano, H., Shim, E., Reed, J. C., Juul, H., Richardson, A. G., de Vries, J., Bink, H., Dichter, M. A., Lucas, T. H., Coulter, D. A., Cubukcu, E., and Litt, B. (2014) Transparent, flexible, low noise graphene electrodes for simultaneous electrophysiology and neuroimaging. Nature Communications, **5**, 5259.
53. Gonzalez-Cuenca, M., Zipprich, W., Boukamp, B. A., Pudmich, G., and Tietz, F. (2001) Impedance studies on chromite-titanate porous electrodes under reducing conditions. *Fuel Cells*, **1**(3-4), 256-264.
54. Oliveira, C. G., and Ferreira, M. G. S. (2003) Ranking high-quality paint systems using EIS. Part I: intact coatings. *Corrosion Science*, **45**(1), 123-138.
55. Zhang, K., Zhang, L. L., Zhao, X. S., and Wu, J. (2010) Graphene/polyaniline nanofiber composites as supercapacitor electrodes. *Chemistry of Materials*, **22**(4), 1392-1401.
56. Das, T. K., and Prusty, S. (2013) Graphene-based polymer composites and their applications. *Polymer-Plastics Technology and Engineering*, **52**(4), 319-331.
57. Wessling, B. (1999) Scientific engineering of anti-corrosion coating systems based on organic metals (polyaniline). *The Journal of Corrosion Science & Engineering*, **1**, paper 15.
58. Lu, W.-K., Elsenbaumer, R. L., and Wessling, B. (1995) Corrosion protection of mild steel by coatings containing polyaniline. *Synthetic Metals*, **71**(1-3), 2163-2166.

8

Inhibition of Corrosion by Polyetherimide-Graphene Nanocomposite Coatings for Long Term Protection of Carbon Steel

8.1 Introduction

Metallic materials exposed to aggressive environment during their application undergo deterioration and premature failure because of the processes of corrosion. Application of organic/polymeric coatings is the standard procedure to prevent the corrosive destruction of metal components. The barrier nature of organic/polymeric coatings prevents the corrosive ions and gases from reaching the metal surface. Recent studies indicate that addition of nanofillers enhance the barrier properties of the coatings [1-4]. Nanofillers improve the adhesion strength between the polymer and metal as well as the gas impermeability, mechanical strength and dimensional stability of the polymer, thus, enhancing the overall corrosion resistance of the coating [2,3,5].

Graphene, one-atom thick two dimensional layers of sp^2 carbon atoms bonded together in a honey comb lattice, has been recently reported as an ideal filler for corrosion inhibition in reinforced polymer coatings because of its unique properties of exceptional chemical and thermal resistance, mechanical strength as well as high electrical and thermal conductivities [6-21]. Graphene prevents penetration of water and corrosive species like oxygen and chlorides due to barrier effect owing to its high aspect ratio and platelet structure which makes the diffusion pathway in the coating tortuous. Graphene produces electron depletion when in direct contact with the metal surface, thus, leading to metal/metal oxide surface passivation through establishment of a Schottky barrier, thereby providing an effective protection of the metal against corrosion by an active-passive approach [6,22].

Several works have been reported on the corrosion inhibitive nature of graphene in polymer composite protective coatings [6-13,23,24]. Most of these demonstrated the diffusion barrier nature of

Tehsin Akhtar, Gisha E. Luckachan and Vikas Mittal, Khalifa University of Science and Technology, Abu Dhabi, UAE
© 2021 Central West Publishing, Australia

graphene, preventing direct interaction between underlying metal and corrosion promoting species. Recently, Chaudhry et al. [23] discussed anti-corrosion behavior of graphene incorporated self-crosslinked poly(vinyl butyral) (PVB) composites. Graphene platelets increased the diffusion path of oxygen and ionic species through the coating which improved corrosion resistance of the coating for a short period of time. For longer periods of time, corrosion promotion occurred due to induced porosity and electrochemical mechanisms related to excellent electrical conductivity of graphene. Figure 8.1 also demonstrates the potentiodynamic curves and optical microscopy images of PVB and graphene composites after 26 h of immersion. Similar mechanisms were reported for graphene/epoxy [11,12],

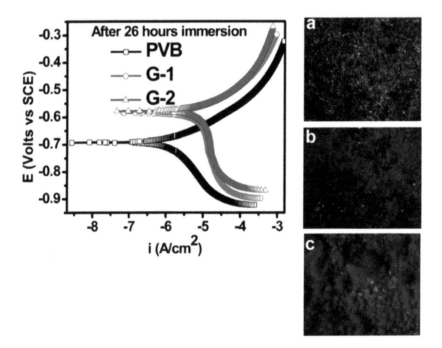

Figure 8.1 Potentiodynamic curve and optical microscopy images of (a) PVB, (b) G-1, (c) G-2 after 26 h of immersion. Reproduced form Reference 23 with permission from Royal Society of Chemistry.

polyurethane/graphene [7,13], polyimide/graphene [10], polyaniline/graphene [9] and poly(sodium styrene sulfonate)/graphene [25] composites in protective coatings. Some researchers also reported that graphene could act as cathode when in contact with metal due to its thermodynamic stability and increase the corrosion due to faster electron transfer owing to high conductivity, thereby, accelerating the metal dissolution and cathodic reduction when the graphene films are non-uniform and non-adherent [16,25,26]. This behavior restricted the protection of graphene grown by chemical vapor deposition (CVD) [26-29], electrophoretic deposition (EPD) [19] and other techniques [18,30] on the metal surface to short durations. Few works have reported the electrochemical response of underlying metal in presence of graphene layers [22,27,28]. Kirkland *et al.* [29] observed that the graphene affected metals differently. It primarily slowed the anodic dissolution reactions for Ni and the cathodic reduction reactions for Cu. Though graphene provided a barrier to metal dissolution of the metal substrate from the environment, however, the mechanism of protection imparted by the graphene layer needs to be elaborated. A quantitative description of p-n junction formed at the metal graphene interface was discussed by Sreevatsa *et al.* [31]. However, the authors used graphene in conjunction with polypyrrole (a conducting polymer) 'top coat' to form the p-n junction. Recently, Dumee *et al.* [22] described the reduced kinetics of corrosion reactions by graphene flakes grown on stainless steel resulted by the electron depletion layer formation which restricted the flow of electrons to the metal surface. Such a behavior of graphene in the protection mechanism was generally not reported in the polymer composite coatings [7-13]. If the electron depletion behavior of graphene could be implemented along with its diffusion barrier nature, the duration of protection by polymer composite coatings could be enhanced significantly. It could be attained by the good dispersion of graphene in the polymer matrix and by the strong interaction of graphene with the polymer chains and metal surface. Therefore, choice of polymer is important in obtaining a better performance of graphene in the polymer composite coatings.

It has been reported that good compatibility exists between polyetherimide (PEI) and graphene as these interact through π–π stacking which is non-covalent in nature and is formed by attractive forces between the adjacent aromatic rings of PEI and graphene [6]. Dennis *et al.* [6] reported such a behavior in graphene incorporated PEI coatings, though, the nature of coating in the aggressive environment as well as the protection mechanism of graphene need to be elaborated further. Therefore, a systematic study is required to further establish the fact that graphene reinforced coatings can serve as effective anti-corrosion alternatives for long-term protection of metals.

The present work demonstrates a one-step procedure for the incorporation of graphene in polyetherimide coating formulations for effective corrosion protection of carbon steel substrates for long periods. Nature of coating in the aggressive environment as well as the mechanism of protection provided by the incorporated graphene platelets was studied by electrochemical analyses and microscopic techniques. Effect of graphene oxide reinforcement instead of graphene platelets in the corrosion protective nature of PEI composite coatings was also analyzed for comparison.

8.2 Experimental

8.2.1 Materials

Carbon steel of grade RST37-2 DIN 17100-80 was purchased from Qatar Steel Industries, Qatar. Its composition in weight% was C (0.125), Mn (0.519), Si (0.016), P (0.014), S (0.005), Al (0.034) and Fe (99.287). Carbon steel coupons of dimensions 5 cm x 2 cm x 1.8 cm were used as substrate. Polyetherimide with melt flow index 9 g/ 10 min (337 °C/6.6 kg) was purchased from Sigma-Aldrich. The graphene nano-platelets (N002-PDR, 98.48% purity) were purchased from Angstron Materials, Dayton, Ohio. N,N-dimethylacetamide (DMAc) solvent and hydrochloric acid (37%) were obtained from Merck and used as received.

8.2.2 Preparation of PEI-Graphene Coatings

Solutions of polyetherimide (15 wt%) were prepared by dissolving PEI pellets in DMAc in closed Pyrex bottles at 70 °C under magnetic stirring to produce homogeneous solutions. Graphene was added to DMAc and mixed under magnetic stirring for 10 min and was subsequently subjected to bath sonication for 6 h and overnight mixing to obtain a homogeneous dispersed graphene solution. This was followed by 2 h of bath sonication and shear mixing for 1 h. The graphene solution was added dropwise to polymer solution prepared separately and stirred vigorously for 30 min, shear mixed for 1 h followed by magnetic stirring overnight to obtain homogeneous solutions [21,32]. The solutions were allowed to stand for 10-15 min to remove air bubbles before dip coating of clean polished carbon steel coupons. Coating was carried out with an immersion speed of 50 mm/min, withdrawal speed of 150 mm/min and wetting time of 30

s. The concentration of graphene was fixed to 0.5, 1, 2 and 5 wt% of the polymer content in the solutions. The coatings were dried immediately in a stepwise manner starting at 80 °C for 30 min and subsequently at 120 °C, 150 °C, 180 °C, 210 °C, 240 °C for 15 min at each temperature and finally curing at 270 °C for 30 min. Graphene-PEI (Gr-PEI) films were also prepared on Teflon petri dishes and subjected to drying in the same manner. 0.5 and 5 wt% coatings with graphene oxide (GO) were also formulated in the same way as Gr-PEI coatings.

8.2.3 Characterization Techniques

The coating thickness was measured using PosiTector 6000 coating thickness gauge. The surface resistance of the coating was measured using the Prostat PRS-812 resistance meter set.

Transmission electron microscopy (TEM) was carried out for gaining insights about the dispersion state of graphene in polymer matrix (Philips CM 20, Germany). The copper grids were immersed for 20 s and quickly withdrawn from the Gr-PEI solutions and dried in the oven at 80 °C for 5 min to obtain a thin coating on copper grids for imaging. Scanning electron microscope (SEM) (FEI Quanta, FEG250, USA) at an accelerating voltage of 5 kV was used for observation of the metal surface underneath the coatings and energy dispersive X-ray spectroscopy (EDX) was employed to confirm the presence/absence of any corrosion products on the metal surface. To study the effect of graphene addition on the decomposition temperature of the coatings, thermogravimetric analysis (TGA) was carried out using the Discovery TGA equipment from TA instruments. The samples were heated from 25 °C to 700 °C at the rate of 10 °C/min under nitrogen flow and the changes in the weights were noted. The Fourier transform infrared spectroscopy of the coatings was performed using a Bruker VERTEX 70 FTIR spectrometer attached with a DRIFT accessory. IR acquisition was achieved by recording 120 scans at a resolution of 4 cm^{-1} in the frequency range of 370 cm^{-1} to 4000 cm^{-1} using OPUS software.

8.2.4 Immersion Tests

The coated coupons were immersed in 3.5 wt% sodium chloride solution and the corrosion was monitored by the appearance of corrosion products or rust with time at room temperature. With the help of optical microscopy (Digital Microscope KH-7700, Hirox Co. Ltd, USA), coated substrates were inspected closely for any signs of corrosion and local delamination. The coatings were scribed after immersion test completion to analyze the extent of undercoating corrosion.

8.2.5 Electrochemical Analysis

A three-electrode cell of 250 mL volume with a platinum gauze counter electrode and a saturated calomel reference electrode (SCE) with bridge tube was used to perform electrochemical tests on the flat coated coupons. The exposed working electrode area was 1 cm^2 and the coated coupon was kept in place with the help of an O ring and adjustable brass disc. All tests were carried out by connecting the corrosion cell to the BioLogic VMP-300 multipotentiostat (controlled by a computer running EC-Lab 10.40 software) in 3.5 wt% NaCl solution at room temperature. Ultra-low current cable connected to the potentiostat was used for the accurate measurement of the current. This option includes current ranges from 100 nA down to 100 pA with additional gains extending the current ranges to 10 pA and 1 pA. The resolution on the lowest range is 76 aA. The open circuit potential (OCP) was measured for 5 min in order to allow the potential to stabilize before the electrochemical impedance and potentiodynamic polarization test. The impedance measurements were performed at amplitude of 15 mV over frequency range from 200 kHz to 10 MHz. Linear polarization was carried out at potential range from -0.250 V to +0.250 V using a scan rate of 1.66 mV/s. Fits were performed by the associated software and corrosion rates were calculated using Faraday's equation for metal loss corrosion. The impedance test and linear polarization was conducted on three samples of each type to ensure reproducibility of the results.

8.3 Results and Discussion

Protective coatings of graphene-polyetherimide composites were prepared by dip coating on carbon steel substrate and subjected to a step wise heating from ambient conditions till 270 °C (as mentioned in the experimental part). Four different Gr-PEI coatings were obtained by changing the concentration of graphene as 0.5, 1, 2 and 5 wt% of PEI in the final coating formulation. Thickness of the obtained coatings was in the range of 20-30 μm.

Graphene in the coatings exhibited a sheet like morphology, as demonstrated in the TEM images shown in Figure 8.2, indicating a uniform distribution of graphene platelets in PEI matrix. The sheet

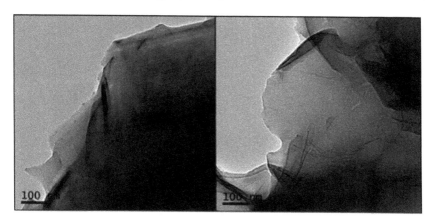

Figure 8.2 TEM images of Gr-PEI films: (left) 0.5 wt% Gr and (right) 5 wt% Gr in PEI.

like morphology added to the flexibility of the Gr-PEI coatings on steel. Further, the formability of the coated samples was also promoted. The sheet like morphology was due to π-π stacking between polyetherimide and graphene sheets [9,28]. Good interfacial compatibility was observed between PEI and graphene also due to similarities in their chemical structure. In addition, the graphene platelets were observed in Figure 8.3 to be effectively stabilized by PEI chains as the Gr-PEI formulations with different concentrations of graphene did not show phase separation even after 4 months of storage. As mentioned above, the uniform dispersion of graphene in PEI matrix was attributed to π-π interaction between aromatic moieties in the polymer and π conjugated graphene basal planes.

The incorporation of graphene also increased the thermal stability of the composite coatings in comparison to the pure PEI by 50 °C. Both 2 wt% and 5 wt% Gr-PEI samples exhibited a delayed decomposition temperature as compared to the pure polymer, as demonstrated in Figure 8.4.

The surface resistance of the coatings was recorded to analyze the changes in the conductivity of polymer coatings due to graphene addition, which can help in further understanding the corrosion behavior. The surface resistance of the pure PEI coating was observed to be high due to the insulating nature, as can be observed from Table 8.1.

Figure 8.3 Gr-PEI formulations after 4 months of storage: (left) 0.5 wt% Gr and (right) 5 wt% Gr.

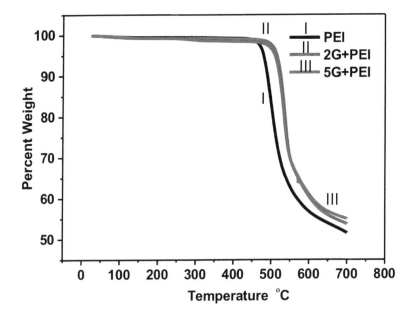

Figure 8.4 TGA thermograms of pure PEI and Gr-PEI films with 2 wt% and 5 wt% graphene.

Table 8.1 Surface conductivities of various coatings

Coating	Surface Resistance Ω	Surface Conductivity S/square
PEI	1.48×10^{11}	6.75×10^{-13}
0.5 wt% Gr-PEI	2.10×10^{11}	4.76×10^{-13}
1.0 wt% Gr-PEI	2.00×10^{11}	5.00×10^{-13}
2.0 wt% Gr- PEI	1.10×10^{4}	9.10×10^{-6}
5.0 wt% Gr-PEI	3.66×10^{2}	2.73×10^{-4}
0.5 wt% GO-PEI	1.10×10^{11}	9.10×10^{-13}
5.0 wt% GO-PEI	4.80×10^{5}	2.08×10^{-7}

A decrease in the resistance of coatings took place upon addition of graphene at concentrations higher than 1 wt%. A sharp reduction in resistance at 2 wt% loading of graphene nano-platelets as compared to 0.5 wt% and 1 wt% loading confirmed the formation of a percolation network structure throughout the polymer matrix [9,21]. At 5 wt% loading of graphene, the resistance declined rapidly to 3.66×10^2 Ω from 1.48×10^{11} Ω, exhibiting a decrease of 9 orders of magnitude in comparison with pure PEI. Meanwhile, the increased conductivity of the coating at 5 wt% graphene indicated a good platelet to platelet connection in forming a conducting path network in the polymer matrix [21,33]. The graphene sheets have high surface area to volume ratio and contact between the well-distributed sheets in the polymer matrix resulted in an increase in the surface conductivity of the coatings.

Protection of metal surface from corrosion by Gr-PEI composite coatings was studied by EIS analysis in 3.5 wt% NaCl solution. The 0.5 wt% and 1 wt% Gr-PEI coatings had similar corrosion performance. The Bode plots in Figures 8.5a and b indicated that the barrier properties of the 0.5 wt% and 1 wt% Gr-PEI coatings declined every week of immersion in NaCl solution. The initial impedance of the 0.5 wt% coating was very high ($10^{11.6}$ Ω.cm^2), however, a gradual decrease in the impedance modulus of the coating occurred and a loss of adhesion and pitting corrosion was observed during 6 weeks of immersion as seen in Figures 8.6a and b. The decline of impedance in 1 wt% Gr-PEI coating was faster than 0.5 wt% Gr-PEI coating and within a week's time of immersion, the coatings exhibited an impedance decline of 3 orders of magnitude, as can be seen from Figure 8.5b. In the subsequent weeks, the impedances decreased further and pitting was observed at several areas of the coating (Figures 8.6c and d). As a result, complete delamination of the coating was observed to

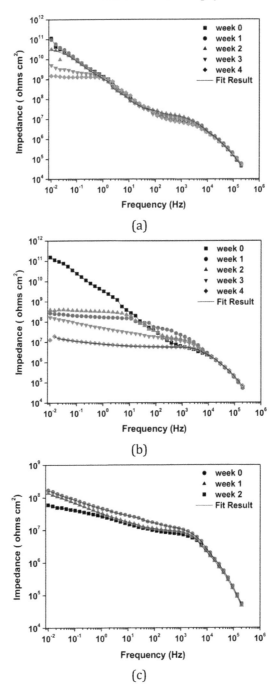

Figure 8.5 continued on next page

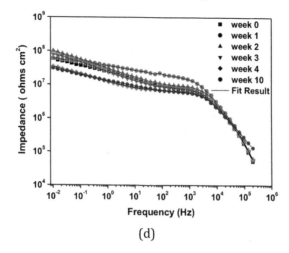

(d)

Figure 8.5 Bode plots of Gr-PEI coatings with graphene concentration of: (a) 0.5 wt% (b) 1 wt% (c) 2 wt% and (d) 5 wt%.

occur during fifth week of immersion. It indicated the diffusion of water into the coating and deterioration of barrier properties. Low frequency impedance decreased continuously to $10^{7.7}$ Ω with increase of graphene concentration to 2 wt% (Figure 8.5c). The coatings also did not achieve the desired level of pore blockage to prevent water from entering the coatings, as it delaminated before the beginning of third week of immersion (Figures 8.6e and f). Interestingly, further increase of graphene to 5 wt% did not decrease low frequency impedance from $10^{7.7}$ Ω and the value was retained over 10 weeks of exposure in 3.5 wt% NaCl solution (Figure 8.5d) without any visible corrosion products on the coating surface (Figures 8.6g and h). The constant coating impedance with immersion time exhibited the stability of adhesion and barrier protection of the 5 wt% Gr-PEI coatings. It can be explained that at low concentrations, the discontinuity of graphene particles created defects in the coatings leading to formation of pockets filled with electrolyte at the interface, and, thus, contributed to a localized corrosion on the metal surface as observed in the 0.5 wt% and 1 wt% graphene coatings (Figures 8.6b and d). At higher concentrations, the dispersed graphene particles, stabilized by the PEI matrix, formed sheet like structures with a good platelet to platelet contact (confirmed from the increased conductivity of the coating), which enhanced the impermeability of the coating because of the hydrophobic and diffusion barrier nature of the graphene platelets [10-12,27]. The metal surface was analyzed by scribing the 5 wt%

230   *Functional Coatings for Corrosion Protection*

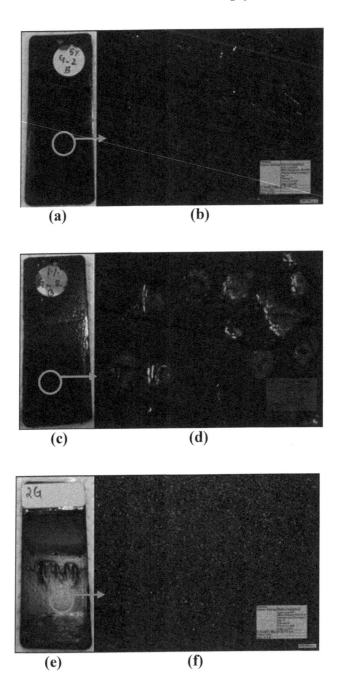

Figure 8.6 continued on next page

Polyetherimide-Graphene Nanocomposite Coatings 231

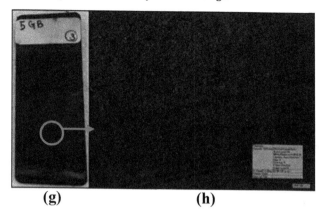

(g)   (h)

Figure 8.6 Digital images of Gr-PEI composite coatings (a) 0.5 wt% Gr, (c) 1 wt% Gr, (e) 2 wt % Gr and (g) 5 wt % Gr after 6 weeks of immersion in 3.5 wt.% sodium chloride solution. (b), (d), (f) and (h) are optical images of the corresponding coatings marked with red circle.

Gr-PEI coating after 10 weeks of exposure to 3.5 wt% NaCl. The optical and scanning electron images of the metal surface at the scribed area did not exhibit any corrosion products (Figures 8.7a and b). It was confirmed further by EDX that observed signals corresponded to only metal ions (Figure 8.7c). In order to analyze the stability of adhesion, a scribed 5 wt% Gr-PEI coating was immersed in 3.5 wt% NaCl over 6 weeks. No coating disbonding was observed for the scribed coating exhibiting that it maintained an excellent adhesion (Figure 8.7d). The coating did not allow the entry of water from the scribed region and lateral diffusion, indicating the good interaction of graphene with PEI matrix and metal surface.

The EIS results were subsequently fitted using the equivalent electric circuit shown in Figure 8.8. R_s represents the solution resistance between the working electrode (coated sample) and the reference saturated calomel electrode. R_c is the coating/pore resistance and Q_c is the constant phase element related to the coating capacitance (C_c) as per the equation

$$C = Y_0 \left(\omega_{max} \right)^{n-1} \qquad (1)$$

where Y_0 is the magnitude of the constant phase element, ω_{max} is the frequency at which the imaginary impedance reaches a maximum for the respective time constant and 'n' is the exponential term of the CPE

232 Functional Coatings for Corrosion Protection

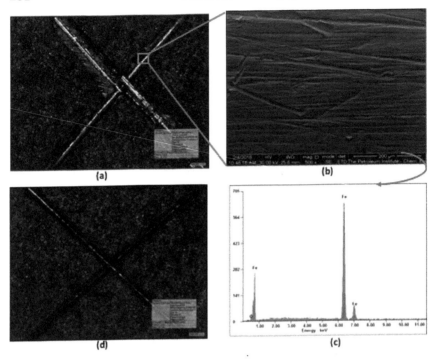

Figure 8.7 Optical image of scribed area on 5 wt% Gr-PEI coating (a). SEM image (b) and EDX (c) of the metal surface on the scribed area marked with red square. Optical image of 5 wt% Gr-PEI coating scribed and immersed for 6 weeks in 3.5 wt% sodium chloride solution.

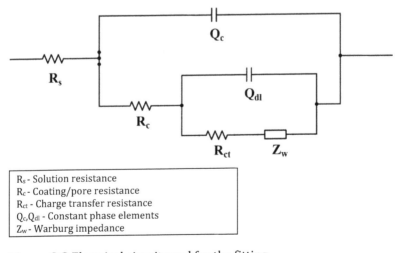

R_s - Solution resistance
R_c - Coating/pore resistance
R_{ct} - Charge transfer resistance
Q_c, Q_{dl} - Constant phase elements
Z_w - Warburg impedance

Figure 8.8 Electrical circuit used for the fitting.

which can vary between 1 for pure capacitance and 0 for a pure resistor [34,35]. Since constant 'n_1' was equal to 1 for all the coatings over the complete measurement time (Table 8.2), Q_c was considered as similar to pure capacitors and, hence, the Y_0 constant followed the same trend as the capacitance [35,36]. Capacitance of a coating (C_c) is

Table 8.2 Fitted parameters from Bode plots of Gr-PEI coatings

Coating	Week	R_{pore} (Ω cm^2)	Q_c		R_{ct} (Ω cm^2)	Q_{dl}		W (Ω s$^{-1/2}$)
			n_1	Y_0 (Ω^{-1}cm^2)		n_2	Y_0 (Ω^{-1}cm^2)	
0.5 wt%	0	6.04x10^6	1	1.04x10^{-11}	5.57x10^9	0.997	5.66x10^{-11}	5.17x10^{10}
	1	4.00x10^6	1	1.12x10^{-11}	3.06x10^9	0.913	1.97x10^{-10}	3.80x10^9
	2	1.99x10^6	1	8.21x10^{-11}	2.98x10^9	0.961	1.61x10^{-10}	9.24x10^7
	3	1.51x10^6	1	8.98x10^{-11}	1.22x10^9	0.953	2.08x10^{-10}	4.00x10^7
	4	1.09x10^6	1	9.37x10^{-11}	5.88x10^7	0.760	9.76x10^{-10}	4.71x10^6
1 wt%	0	6.42x10^6	1	9.67x10^{-12}	1.25x10^9	0.986	4.80x10^{-11}	8.48x10^{10}
	1	4.77x10^6	1	9.84x10^{-12}	1.14x10^8	0.903	1.38x10^{-10}	1.91x10^7
	2	1.90x10^6	1	9.78x10^{-12}	3.05x10^8	0.941	9.83x10^{-11}	1.42x10^7
	3	1.11x10^6	1	9.90x10^{-12}	2.66x10^8	0.379	1.19 x 10^{-8}	1.57x10^7
	4	9.52x10^5	1	1.05x10^{-11}	1.39x10^7	0.476	1.33 x 10^{-7}	3.41x10^5
2 wt%	0	7.63x10^6	1	9.43x10^{-12}	4.08x10^7	0.529	1.16x10^{-8}	3.25x10^6
	1	6.12x10^6	1	9.33x10^{-12}	1.76x10^7	0.407	1.06x10^{-8}	2.47x10^7
	2	4.20x10^6	1	1.06x10^{-11}	1.42x10^7	0.496	1.39x10^{-8}	1.3x10^7
5 wt%	0	7.21x10^6	1	9.93x10^{-12}	3.74x10^7	0.506	1.34x10^{-8}	-
	1	8.73x10^6	1	8.47x10^{-12}	3.74x10^7	0.513	1.14x10^{-8}	-
	2	8.63x10^6	1	8.35x10^{-12}	9.94x10^7	0.577	1.49x10^{-8}	-
	3	7.24x10^6	1	9.39x10^{-12}	8.71x10^7	0.589	1.95x10^{-8}	-
	4	7.17x10^6	1	9.11x10^{-12}	7.65x10^7	0.583	1.68x10^{-8}	-
	10	6.93x10^6	1	9.90x10^{-12}	6.90x10^7	0.618	1.66x10^{-8}	-

proportional to its dielectric constant and can be, therefore, attributed to the amount of water absorbed by the coating [36]. Therefore, the changes in Y_0 constant of Q_c with immersion time of the coatings (mentioned in Table 8.2) provides insights about the tendency of coatings to water and ionic species. For 0.5, 1 and 2 wt% Gr coatings, Y_0 constant increased significantly at the end of the measurement time. On the other hand, for 5 wt% Gr coating, the final value remained in the same order of 10^{-12} observed in the beginning of the immersion test, thus, exhibiting lower permeability of coatings at high graphene content. It indicated that the well dispersed graphene platelets provided diffusion barrier by enhancing both hydrophobicity of the coating and the tortuous paths of corrosive agents through the coating, thus, greatly reducing the corrosion tendency of underlying metal. It was also observed further in the R_{ct} data provided in

Table 8.2. R_{ct} is attributed to the resistance to charge transfer processes on the metal surface. Addition of graphene particles diminished R_{ct} in the following order of 0.5>1>2>5 wt% graphene in PEI matrix. Except 5 wt% coating, all other coatings exhibited a large decrease in R_{ct} value during a 3-week immersion. However, a different behavior was observed for 5 wt% coating, where R_{ct} value increased after 2 weeks and the final R_{ct} value obtained after 10 weeks was higher than the initial value.

The behavior of Gr-PEI coatings was further analyzed by potentiodynamic measurements. Figures 8.9 a to c present the anodic and cathodic polarization curves of Gr-PEI coatings in 3.5 wt% NaCl solution. The corresponding E_{corr} and I_{corr} obtained by the fitting analysis are plotted in Figures 8.9 e to g. The polarization curves of 0.5 wt% Gr-PEI and 2 wt% Gr-PEI coatings in Figures 8.9 a and b showed the decline of coatings' barrier properties due to shifts of the curves from noble positive potentials to negative potentials with increase in immersion time. 2 wt% Gr-PEI coatings delaminated very fast during polarization testing and, therefore, accurate readings after 2 weeks of immersion could not be recorded. The polarization plots of 5 wt% Gr-PEI coatings in Figure 8.9 c showed an interesting behavior. A shift in the E_{corr} towards more noble or positive value (-117.728 mV) was recorded as compared with the initial value of -225.264 mV during immersion. After 4 weeks of immersion, the E_{corr} shifted further towards positive potential values (-53.960 mV). For comparison, the potentiodynamic curves of pure PEI coating were also plotted in Figure 8.9 d. Since E_{corr}, cathodic and anodic currents or slopes were absent in the initial weeks of immersion, I_{corr} of the pure PEI coating during this time was calculated directly from the Tafel plots without performing any fitting analysis [37]. In the beginning, E_{corr} value of pure PEI coating was higher than 5 wt% Gr-PEI coating. However, it decreased gradually, as expected from a barrier coating, to -185.88 mV after 4 weeks of immersion which was more negative than the E_{corr} value of 5 wt% Gr-PEI coating (-53.960 mV). Thus, all coatings, except with 5 wt% Gr, exhibited a shift of E_{corr} towards more negative potentials with time indicating that the protection provided by the coatings was purely barrier in nature, which could degrade over time. I_{corr} plotted in Figures 8.9 e to h showed a continuous increase with immersion time for pure PEI and 0.5 wt% Gr-PEI coatings. I_{corr} decreased continuously for 5 wt% Gr-PEI coatings, though, towards the end of the measurement it increased slightly, which can be accounted to the enhanced electrical conductivity of the coating due to the high graphene

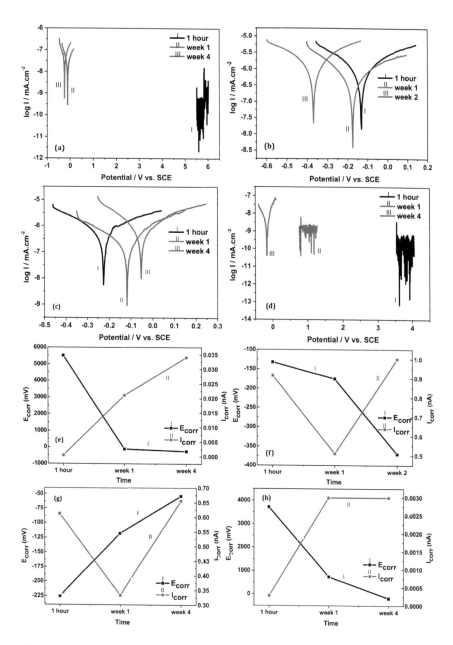

Figure 8.9 Tafel plots (a-d) and E_{corr} & I_{corr} vs. time of immersion (e-h) of Gr-PEI coatings with 0.5 wt% (a) & (e), 2 wt% (b) & (f), 5 wt% (c) & (g) graphene respectively. (d) & (h) represent the characteristics of pure PEI respectively.

content. The shift of the corrosion potential towards the positive scale only for 5 wt% coatings confirmed that graphene inhibited the corrosion of steel substrate at high concentrations when a substantial amount of graphene was in contact with the metal at the interface of the coating and metal. Similar shift was observed by Dennis *et al.* [6] in a Gr-PEI system with 2 wt% and 20 wt% graphene loadings. A corrosion rate of 5.52×10^{-4} mmpy and 8.46×10^{-4} mmpy respectively was reported for these coatings which is higher than that of 5 wt% Gr-PEI coating (3.9×10^{-6} mmpy) prepared in the current study.

A pictorial representation of the protection provided by the graphene in PEI matrix is presented in Figure 8.10. During the initial hours of immersion in NaCl solution, the 5 wt% Gr-PEI coating prevented water and oxygen permeation through the coating due its hydrophobic nature and platelet structure with high aspect ratio which increased the diffusion pathway of the corrosive species and inhibited the corrosion reaction by barrier effect. However, at prolonged

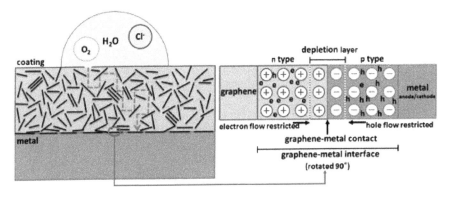

Figure 8.10 Pictorial representation of protection provided by graphene in 5 wt% Gr-PEI composite coating.

immersion, the protection exhibited by 5 wt% Gr-PEI coating was obtained by a different mechanism. Graphene promoted steel passivation at the interface of the steel surface and coating. This noble behavior of steel might have occurred due to the formation of Schottky barrier at the steel-graphene junction [22]. The insert in Figure 8.10 shows the protection obtained by the band gap formation at the metal-graphene interface. The graphene is n-doped at the interface when contact with metal is established [38-41]. It is reported that adsorption of graphene on the metal includes initial charge transfer and hybridization [42-44]. Thus, the steel-graphene interface forms a p-n junction with metal surface as the p-type semiconductor and graphene as the n-type semiconductor (Figure 8.10). The depletion region or potential barrier does not allow movement of major charge carriers, i.e., electrons (e), from graphene to the metal surface and holes (h) from the metal surface to graphene. Electron depletion occurs at the metal substrate (anode) and no electrons are available for transfer to the cathode. This reduces the tendency of the metal to oxidize. The metal acts more nobly due to the passivation by the contact with graphene. In addition, the electron transfer from graphene coating to the metal (cathode) is prevented by the presence of potential barrier/depletion region. This also means that had there been electrons at the anode available for cathodic reaction, the movement of electrons from the anode through graphene to the cathode would not have been possible. Therefore, the cathodic reduction of oxygen is also arrested by the formation of p-n junction. Moreover, the repelling field created by the concentration of electrons in the p orbitals of graphene can repel negatively charged ions like chloride and improve anti-corrosion performance [45]. Such a protection was not reported in many graphene based coatings reported in the literature, where graphene provided protection for short time durations by increasing the tortuous path of diffusion through the coating. The passivation of steel by the n-doping of graphene at the metal interface provided protection for long time durations.

The surface area of the steel covered by graphene had an impact on the corrosion kinetics. The 0.5, 1 and 2 wt% coatings did not provide anodic protection of steel due to low concentration of graphene, as fuller coverage of the steel surface was not achieved. This exhibited that graphene coatings did not provide good anti-corrosion performance if these were not adsorbed on the metal at sufficiently high density [15,46]. Thus, any pathway leading to the unbound graphene can cause poor local adhesion as already observed for 0.5, 1, 2 wt% coatings. However, this was not the case for 5 wt% Gr coatings, which achieved excellent corrosion resistance for long periods because of the good contact between graphene and metal. It was achieved by the strong interaction of PEI with graphene and metal surface as observed from the FTIR spectra presented in Figure 8.11. Compared with the IR spectra of PEI pellet, the pure PEI coating exhibited two new signals at 1708 cm^{-1} and 1343 cm^{-1} corresponding respectively to changes in the imide carbonyl and C-N-C bonds (Figures 8.11 a & b) [47]. This observation indicated that the functional groups of imide ring in polyetherimide, C=O as well as C-N-C groups, interacted with

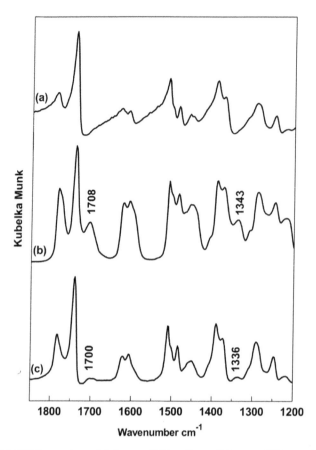

Figure 8.11 FTIR spectra of (a) pure PEI pellets, (b) pure PEI coating and (c) 5 wt% Gr-PEI composite coating.

the steel extensively to form strong adhesive bonds. These signals appeared in the IR spectra of 5 wt% Gr-PEI coating at low wavenumbers of 1700 cm^{-1} and 1336 cm^{-1} respectively. This might have occurred due to the interaction of PEI chains with metal surface as well as the graphene platelets. These strong adhesive forces stabilized the graphene platelets on the metal surface and, thus, enabled them to involve in the passivation process of steel by the formation of p-n junction. Therefore, strong binders like PEI are required as a matrix to maintain a good adhesion to the steel substrate which would help graphene particles to provide protection for long term.

Graphene oxide (GO) based coatings with filler concentrations of 0.5 and 5 wt% were also generated for comparing the performance with graphene based coatings. The Bode plots for these coatings are presented in Figure 8.12. It can be observed that the coatings with GO were unstable and exhibited decline in impedance values just after a week of immersion. Pitting corrosion and fast delamination occurred in 5 wt% GO-PEI coatings. The initial impedance of this coating was lower than that of 5 wt% Gr-PEI coating and further decreased significantly within a week of immersion. This demonstrated that GO did not protect the steel from corrosion and failed to provide an effective barrier against redox reactions. The corrosive electrolyte penetrated through the coatings and the coating performance deteriorated due to hydrophilic nature of GO as well as lower conductivity of the 5 wt% GO-PEI coating. For instance, the surface resistance of 5 wt% GO-PEI coating (4.80×10^5 Ω) was 3 orders higher than 5 wt% Gr-PEI coating (3.66×10^2 Ω), thus, indicating the lower conductivity of PEI matrix with GO incorporation.

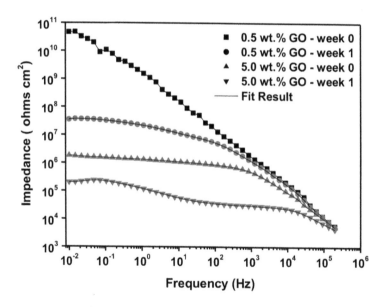

Figure 8.12 Bode plots of GO-PEI composite coatings.

8.4 Conclusions

The present work aimed to synthesize graphene incorporated stable polymer composite coatings for long term protection of carbon steel. Graphene (Gr) was incorporated in polyetherimide (PEI) matrix in different concentrations to investigate the anti-corrosion effect of graphene and nature of protection offered by Gr-PEI composite coatings. A maximum of 5 wt% graphene was successfully incorporated in the polyetherimide matrix, without sacrificing its native barrier property, by using a heating cycle starting from ambient temperature to 270 °C. Graphene loading of 5 wt% maintained good adhesion of the coating, while the coatings with other graphene concentrations (0.5, 1 and 2 wt%) failed due to their inability to stop the influx of water due to low concentration of graphene platelets as well as stresses which destroyed adhesive forces. The low frequency impedance modulus decreased continuously for 0.5, 1 and 2 wt% Gr-PEI coatings, whereas it remained constant at $10^{7.7}$ Ω over 10 weeks of exposure in 3.5 wt% NaCl solution without any visible corrosion products on the coating surface for 5 wt% Gr-PEI coating. During the initial hours of immersion, the 5 wt% Gr-PEI coating prevented water and oxygen permeation due to its hydrophobic nature and graphene platelets with high aspect ratio which increased the diffusion pathway of the corrosive species and inhibited the corrosion reaction by barrier effect. At prolonged immersion, the coating worked through electron depletion at the surface of steel by the formation of a p-n junction, which suppressed the metal's ability to donate electrons and oxidize. The potential barrier at the p-n junction also stopped electron transfer from the coating to the cathode. Thus, both anodic and cathodic reactions were stopped by the addition of graphene at high concentrations. It was also confirmed by an increased R_{ct} and a positively shifted E_{corr} of 5 wt% Gr-PEI composite coatings after a long period of 4-week immersion. Therefore, it can be suggested that graphene can provide a long-term protection when incorporated with suitable polymer matrices, thereby, generating protective barrier coatings with enhanced lifespan.

References

1. Behzadnasab, M., Mirabedini, S., and Esfandeh, M. (2013) Corrosion protection of steel by epoxy nanocomposite coatings containing various combinations of clay and nanoparticulate zirconia. *Corrosion Science*, **75**, 134-141.
2. Zhang, Y., Shao, Y., Zhang, T., Meng, G., and Wang, F. (2013) High corrosion protection of a polyaniline/organophilic montmorillonite coating for magnesium alloys. *Progress in Organic Coatings*, **76**, 804-811.
3. Piromruen, P., Kongparakul, S., and Prasassarakich, P. (2014) Synt-

hesis of polyaniline/montmorillonite nanocomposites with an enhanced anticorrosive performance. *Progress in Organic Coatings*, **77**, 691-700.

4. Radhakrishnan, S., Siju, C., Mahanta, D., Patil, S., and Madras, G. (2009) Conducting polyaniline–nano-TiO2 composites for smart corrosion resistant coatings. *Electrochimica Acta*, **54**, 1249-1254.

5. Bagherzadeh, M., and Mousavinejad, T. (2012) Preparation and investigation of anticorrosion properties of the water-based epoxy-clay nanocoating modified by Na+-MMT and Cloisite 30B. *Progress in Organic Coatings*, **74** 589-595.

6. Dennis, R. V., Viyannalage, L. T., Gaikwad, A. V., Rout, T. K., and Banerjee, S. (2013) Graphene nanocomposite coatings for protecting low-alloy steels from corrosion. *American Ceramic Society Bulletin*, **92**(5), 18-24.

7. Li, Y., Yang, Z., Qiu, H., Dai, Y., Zheng, Q., Li, J., and Yang, J. (2014) Self-aligned graphene as anticorrosive barrier in waterborne polyurethane composite coatings. *Journal of Materials Chemistry A*, **2**, 14139-14145.

8. Chang, K.-C., Hsu, M.-H., Lu, H.-I., Lai, M.-C., Liu, P.-J., Hsu, C.-H., Ji, W.-F., Chuang, T.-L., Wei, Y., and Yeh, J.-M. (2014) Room-temperature cured hydrophobic epoxy/graphene composites as corrosion inhibitor for cold-rolled steel. *Carbon*, **66**, 144-153.

9. Chang, C.-H., Huang, T.-C., Peng, C.-W., Yeh, T.-C., Lu, H.-I., Hung, W.-I., Weng, C.-J., Yang, T.-I., and Yeh, J.-M. (2012) Novel anticorrosion coatings prepared from polyaniline/graphene composites. *Carbon*, **50**(14), 5044-5051.

10. Chang, K., Hsu, C., Lu, H., Ji, W., Chang, C., Li, W., Chuang, T., Yeh, J., Liu, W., and Tsai, M. (2014) Advanced anticorrosive coatings prepared from electroactive polyimide/graphene nanocomposites with synergistic effects of redox catalytic capability and gas barrier properties. *Express Polymer Letters*, **8**(4), 243-255.

11. Liu, S., Gu, L., Zhao, H., Chen, J., Yu, H. (2016). Corrosion resistance of graphene-reinforced waterborne epoxy coatings. *Journal of Materials Science and Technology*, **32**(5), 425-431.

12. Zhang, Z., Zhang, W., Li, D., Sun, Y., Wang, Z., Hou, C., Chen, L., Cao, Y., and Liu, Y. (2015) Mechanical and anticorrosive properties of graphene/epoxy resin composites coating prepared by in-situ method. *International Journal of Molecular Ssciences*, **16**(1), 2239-2251.

13. Mo, M., Zhao, W., Chen, Z., Yu, Q., Zeng, Z., Wu, X., and Xue, Q. (2015) Excellent tribological and anti-corrosion performance of polyurethane composite coatings reinforced with functionalized graphene and graphene oxide nanosheets. *RSC Advances*, **5**, 56486-56497.

14. Sahu, S. C., Samantara, A. K., Seth, M., Parwaiz, S., Singh, B. P., Rath,

P. C., and Jena B. K. (2013) A facile electrochemical approach for development of highly corrosion protective coatings using graphene nanosheets. *Electrochemistry Communications*, **32**, 22-26.

15. Dong, Y., Liu, Q., and Zhou, Q. (2015) Time-dependent protection of ground and polished Cu using graphene film. *Corrosion Science*, **90**, 69-75.

16. Sun, W., Wang, L., Wu, T., Wang, M., Yang, Z., Pan, Y., and Liu, G. (2015) Inhibiting the corrosion-promotion activity of graphene. *Chemistry of Materials*, **27**(7), 2367-2373.

17. Miskovic-Stankovic, V., Jevremovic, I., Jung, I., and Rhee, K. (2014) Electrochemical study of corrosion behavior of graphene coatings on copper and aluminum in a chloride solution. *Carbon*, **75**, 335-344.

18. Liu, J., Hua, L., Li, S., and Yu, M. (2015) Graphene dip coatings: An effective anticorrosion barrier on aluminum. *Applied Surface Science*, **327**, 241-245.

19. Singh, B. P., Nayak, S., Nanda, K. K., Jena, B. K., Bhattacharjee, S., and Besra, L. (2013) The production of a corrosion resistant graphene reinforced composite coating on copper by electrophoretic deposition. *Carbon*, **61**, 47-56.

20. Ming, H., Wang, J., Zhang, Z., Wang, S., Han, E.-H., and Ke, W. (2014) Multilayer graphene: a potential anti-oxidation barrier in simulated primary water. *Journal of Materials Science and Technology*, **30**(11), 1084-1087.

21. Kumar, S., Sun, L., Caceres, S., Li, B., Wood, W., Perugini, A., Maguire, R., and Zhong, W. (2010) Dynamic synergy of graphitic nanoplatelets and multi-walled carbon nanotubes in polyetherimide nanocomposites. *Nanotechnology*, **21**(10), 105702.

22. Dumee, L. F., He, L., Wang, Z., Sheath, P., Xiong, J., Feng, C., Tan, M. Y., She, F., Duke, M., and Gray, S. (2015) Growth of nano-textured graphene coatings across highly porous stainless steel supports towards corrosion resistant coatings. *Carbon*, **87**, 395-408.

23. Chaudhry, A., Mittal, V., and Mishra, B. (2015) Inhibition and promotion of electrochemical reactions by graphene in organic coatings. *RSC Advances*, **5**, 80365-80368.

24. Mayavan, S., Siva, T., and Sathiyanarayanan, S. (2013) Graphene ink as a corrosion inhibiting blanket for iron in an aggressive chloride environment. *RSC Advances*, **3**(47), 24868-24871.

25. Hsieh, Y.-P., Hofmann, M., Chang, K.-W., Jhu, J. G., Li, Y.-Y., Chen, K. Y., Yang, C. C., Chang, W.-S., and Chen, L.-C. (2013) Complete corrosion inhibition through graphene defect passivation. *ACS Nano*, **8**(1), 443-448.

26. Schriver, M., Regan, W., Gannett, W. J., Zaniewski, A. M., Crommie, M. F., and Zettl, A. (2013) Graphene as a long-term metal oxidation barrier: worse than nothing. *ACS Nano*, **7**(7), 5763-5768.

27. Raman, R. S., Banerjee, P. C., Lobo, D. E., Gullapalli, H., Sumandasa, M., Kumar, A., Choudhary, L., Tkacz R., Ajayan P. M., and Majumder, M. (2012) Protecting copper from electrochemical degradation by graphene coating. *Carbon*, **50**(11), 4040-4045.
28. Prasai, D., Tuberquia, J. C., Harl, R. R., Jennings, G. K., and Bolotin, K. I. (2012) Graphene: corrosion-inhibiting coating. *ACS Nano*, **6**(2), 1102-1108.
29. Kirkland, N., Schiller, T., Medhekar, N., and Birbilis, N. (2012) Exploring graphene as a corrosion protection barrier. *Corrosion Science*, **56**, 1-4.
30. Nilsson, L., Andersen, M., Balog, R., Lægsgaard, E., Hofmann, P., Besenbacher, F., Hammer, B., Stensgaard, I., and Hornekær, L. (2012) Graphene coatings: probing the limits of the one atom thick protection layer. *ACS Nano*, **6**(11), 10258-10266.
31. Sreevatsa, S., Banerjee, A., and Haim, G. (2009) Graphene as a permeable ionic barrier. *ECS Transactions*, **19**(5), 259-264.
32. Longun J., and Iroh, J. (2012) Nano-graphene/polyimide composites with extremely high rubbery plateau modulus. *Carbon*, **50**(5), 1823-1832.
33. Luong, N. D., Hippi, U., Korhonen, J. T., Soininen, A. J., Ruokolainen, J., Johansson, L.-S., Nam, J.-D., and Seppala, J. (2011) Enhanced mechanical and electrical properties of polyimide film by graphene sheets via in situ polymerization. *Polymer*, **52**(23), 5237-5242.
34. Zheludkevich, M., Serra, R., Montemor, M., Yasakau, K., Salvado, I. M., and Ferreira, M. (2005) Nanostructured sol–gel coatings doped with cerium nitrate as pre-treatments for AA2024-T3: corrosion protection performance. *Electrochimica Acta*, **51**(2), 208-217.
35. Hsu, C. H., and Mansfeld, F. (2001) Technical Note: Concerning the conversion of the constant phase element parameter Y0 into a capacitance. *Corrosion*, **57**(9), 747-748.
36. Moreno, C., Hernandez, S., Santana, J., Gonzalez-Guzman, J., Souto, R., and Gonzalez, S. (2012) Characterization of water uptake by organic coatings used for the corrosion protection of steel as determined from capacitance measurements. *International Journal of Electrochemical Science*, **7**, 7390-7403
37. Rout, T. K., and Gaikwad, A. V. (2015) In-situ generation and application of nanocomposites on steel surface for anti-corrosion coating. *Progress in Organic Coatings*, **79**, 98-105.
38. Gong, C., Lee, G., Shan, B., Vogel, E. M., Wallace, R. M., and Cho, K. (2010) First-principles study of metal–graphene interfaces. *Journal of Applied Physics*, **108**, 123711.
39. Sevincli, H., Topsakal, M., Durgun, E., and Ciraci, S. (2008) Electronic and magnetic properties of 3d transition-metal atom adsorbed graphene and graphene nanoribbons. *Physical Review B*, **77**, 195434.

40. Enciu, D., Nemnes, G.A., and Ursu, I. (2014) Spintronic devices based on graphene nanoribbons with transition metal impurities. Towards space applications. *INCAS Bulletin*, **6**(1), 45.
41. Jaiswal, N. K., and Srivastava, P. (2013) Fe-doped armchair graphene nanoribbons for spintronic/interconnect applications. *IEEE Transactions on Nanotechnology*, **12**(5) 685-691.
42. Kim, S., Shin, D. H., Kim, C. O., Kang, S. S., Kim, J. M., Jang, C. W., Joo, S. S., Lee, J. S., Kim, J. H., and Choi, S.-H. (2013) Graphene p–n vertical tunneling diodes. *ACS Nano*, **7**(6), 5168-5174.
43. Batzill, M. (2012) The surface science of graphene: Metal interfaces, CVD synthesis, nanoribbons, chemical modifications, and defects. *Surface Science Reports*, **67**(3-4), 83-115.
44. Gadipelli, S., and Guo, Z. X. (2015) Graphene-based materials: synthesis and gas sorption, storage and separation. *Progress in Materials Science*, **69**, 1-60.
45. Berry, V. (2013) Impermeability of graphene and its applications. *Carbon*, **62**, 1-10.
46. Sun, M., Feng, J., Bu, Y., Wang, X., Duan, H., and Luo, C. (2015) Graphene coating bonded onto stainless steel wire as a solid-phase microextraction fiber *Talanta*, **134**, 200-205.
47. da Conceicao, T. F., Scharnagl, N., Dietzel, W., and Kainer, K. (2011) Corrosion protection of magnesium AZ31 alloy using poly(ether imide) [PEI] coatings prepared by the dip coating method: Influence of solvent and substrate pre-treatment. *Corrosion Science*, **53**(1), 338-346.

9

Effect of GO and rGO on the Corrosion Resistance of PMMA Nanocomposite Coatings

9.1 Introduction

The destructive process of metal corrosion caused an annual global loss of $350 million in 2010, which grew to an estimated cost of $900 million in 2015 [1,2]. Widely used polymeric coatings for the protection of metal structures are not effective when defects form in them due to the environmental effects. In fact, no polymer is an absolute barrier against water vapor, gases and organic substances. The application of nanomaterials combined with suitable polymer chemistry paved the way for the development of next-generation coatings with enhanced barrier properties [3]. Nanomaterials improve the adhesion strength between the polymer and metal substrate as well as the gas impermeability and mechanical performance of the polymer, thus, enhancing the corrosion resistance of the coating [4-7]. Recently, graphene based materials like graphene (G), graphene oxide (GO) and reduced graphene oxide (rGO) were employed as ideal nanofillers for corrosion inhibition in reinforced polymer coatings because of exceptional chemical and thermal resistance, mechanical strength as well as high electrical and thermal conductivities of graphene based materials [8-25]. Nanolayered filler platelets provide an extraordinarily torturous diffusion path to the permeant molecules, thus, leading to enhanced barrier performance against gas, moisture and oxygen transmission. In addition to diffusion, the solubility of the corrosive species in the polymer nanocomposites can be markedly influenced by the presence of graphene particles. Thus, the incorporation of the graphene platelets can affect the overall permeability of the corrosive molecules through the polymer coatings or membranes.

The promising approach of incorporating graphene in the polymer matrices for enhanced corrosion protection has been reported in several studies, though, the behavior of graphene derivatives, such as graphene oxide and reduced graphene oxide, has not been reported

Gisha Elizabeth Luckachan, Khaled Hassan and Vikas Mittal, Khalifa University of Science and Technology, Abu Dhabi, UAE
© 2021 Central West Publishing, Australia

in detail. The property of GO to adhere to the surfaces and its function as a surfactant, along with the lamellar surface laden with oxygen functionalities, are desirable components of the coatings to suppress the corrosion of metal components [26,27]. Although the electrical and thermal properties of rGO are not comparable to those of pristine graphene due to structural defects, there are still many advantages in real-life applications. Yoo *et al.* [26] reviewed the barrier applications of graphene and graphene oxide reinforced polymer composites. It was reported that the permeability of oxygen [28-30], carbon dioxide [29], nitrogen [31] moisture [32], helium [33] and hydrogen [34] in polymer composite films could be significantly reduced with the addition of graphene and its derivatives. Krishnamoorthy *et al.* [1] also developed a multifunctional graphene oxide (GO) nano-paint by incorporating GO sheets in an alkyd resin with suitable non-toxic additives using ball milling. The GO nano-paint coating exhibited a corrosion protection efficiency of about 76% in salt water and inhibited the bacterial growth on the surface. GO has also been functionalized to obtain polyurethane acrylate (PUA)/GO composites, which upon UV curing exhibited improved thermal stability and mechanical properties [35]. With 1 wt% functionalized GO, the degradation temperature increased from 299 °C for pure PUA to 316 °C for the composite. The storage modulus for the 1 wt% composite increased by 37% and the tensile strength increased by 73% (relative to pure PUA).

In this work, a comparison of the role of GO and rGO in protecting the metal corrosion was carried out using poly(methyl methacrylate) (PMMA) as the coating matrix. The role of solvents in fabricating suitable barrier coatings as well as the effect of probe sonication time on rGO and GO dispersion in PMMA formulations were also studied.

9.2 Experimental

9.2.1 Materials

PMMA was purchased from Sigma Aldrich and was used as received. The rGO nanofiller was purchased from Angstron Materials, USA (grade N008-100-N). Graphene oxide was synthesized in the laboratory via modified Hummer's method. Carbon steel of grade RST37-2 DIN 17100-80 was purchased from Qatar Steel Industries Factory. Its composition in wt% was C (0.125), Mn (0.519), Si (0.016), P (0.014), S (0.005), Al (0.034) and Fe (99.287). Reagent grade fuming hydrochloric acid (35%) was supplied by Merck. N,N-dimethyl acetamide

(DMA), N-methylpyrrolidone (NMP), toluene, chloroform and acetone were purchased from Sigma Aldrich.

9.2.2 Substrate Preparation

Carbon steel coupons of 5 cm x 2 cm x 2 cm dimensions were pickled with hydrochloric acid for 2 h in order to remove the oxide layer from the surface. After acid treatment, the coupons were polished with sandpaper (60, 150 and 180 grits), followed by rinsing with water and acetone. Finally, the coupons were sonicated in acetone for 10 min and dried in oven at 90 °C for 1 h.

9.2.3 Preparation of Coatings

For selecting a suitable solvent for preparing the nanocomposite coatings, 2.5 g of PMMA was dissolved in 10 mL of DMA, NMP, toluene and chloroform. The polymer was dissolved under stirring and heating until a homogeneous and clear solution was obtained. For low boiling point solvents, a small amount of solvent evaporated and was subsequently replenished to keep the overall amount of solvent as 10 mL. The polymer solution was degassed for 1 h before applying to the polished steel substrates by dip coating. The dip coating parameters were: withdrawal rate 140 mm/min, immersion rate 90 mm/min and immersion time of 2 min. The coated substrates were kept in an atmospheric oven overnight at a temperature 10-15 °C below the boiling point of the chosen solvent. The dip coating procedure was repeated the following day using the same parameters. At the end of the second overnight drying cycle, the coated substrates were further subjected to 270 °C for 45 min.

For nanocomposite coatings, only DMA and NMP were used. The filler loadings of 1, 3, and 5 wt% based on the polymer weight were used. Reduced graphene oxide was added to 10 mL of DMA (or NMP) and stirred for 2 h, followed by probe sonication. Probe sonication (model CPX 750 supplied by Cole Parmer) for each solvent was performed for two different durations, 25 mins and 60 min with pulse intervals (5 sec on, 5 sec off). During probe sonication, the samples were jacketed by an ice bath to minimize the solvent loss. For GO based formulations, GO was stirred in DMA for 2 h and sunsequently bath sonicated for 6 h. After sonication, PMMA pellets were added batch wise to the rGO and GO dispersions over a stirrer plate until the whole amount of polymer dissolved. Finally, the solutions were shear mixed for 1 h to ensure delamination and mixing of the rGO and GO sheets in the polymer. The solutions were applied using dip coating with same coating parameters as pre PMMA coatings. Free-standing films of the materials were also developed by casting the formulations in Teflon molds and employing the same drying conditions as coatings.

9.2.4 Characterization Techniques

The thickness of the coatings was measured using PosiTector 6000 coating thickness gage from DeFelsko Corporation, USA. Tape adhesion tests were carried out on the coated substrates as per ASTM D-3359 standard. The adhesion toolkit used for the purpose was supplied by GARDCO®, and a carbide knife was used to scribe the coatings. A cross-cut was created (test method B) followed by the application of a pressure sensitive tape over the scribed area. The grid pattern resulting on the tape was used to evaluate the adhesion quality as per the standard. The electrical conductivity of the coatings was measured using Prostat PRS-812 meter. Thermogravimetric analysis (TGA) of the prepared PMMA films was carried out using Discovery TGA supplied by TA Instruments. The PMMA films were heated from 25 °C to 500 °C at 10 °C /min heating rate under N_2 purge. The differential scanning calorimetry (DSC) technique revealed the thermal transitions associated with the polymer nanocomposites (PNCs) such as glass transition temperature (T_g) and peak melting temperature (T_m). DSC was carried out using Discovery DSC supplied by TA Instruments. The samples were tested using two heating and cooling cycles (25 °C ↔ 250 °C) with 10 °C/min heating/cooling rate under N_2 purge. Fourier transform infrared (FTIR) spectroscopy of the coatings was performed using a Bruker VERTEX 70 FTIR spectrometer attached with a DRIFT accessory. IR acquisition was achieved by 120 scans at a resolution of 4 cm^{-1} in the frequency range of 370 cm^{-1} to 4000 cm^{-1} using the OPUS software. Wide-angle X-ray diffraction (WAXRD) was performed to evaluate the layered structure of rGO and GO. WAXRD was also performed on the specimen films to investigate the structural changes.

9.2.5 Electrochemical Analysis

A three electrode cell of 250 mL volume with a platinum gauze counter electrode and a saturated calomel reference electrode (SCE) with bridge tube was used to perform electrochemical tests on the flat coated coupons. The coated coupons acted as the working electrode, and an area of 1 cm² was exposed to 3.5 wt% NaCl solution. All tests were carried out at room temperature by connecting the corrosion cell to the BioLogic VMP-300 multi-potentiostat (controlled by a computer running EC-Lab 10.40 software). Ultra-low current cables connected to the potentiostat were used for the accurate measurement of the current. This option includes current ranges from 100 nA down to 100 pA with additional gains extending the current ranges to 10 pA and 1 pA. The resolution on the lowest range was 76 aA. The open circuit potential (OCP) was measured for 5 min in order to allow the potential to stabilize before the electrochemical impedance and potentiodynamic polarization tests. The impedance measurements were performed at an amplitude of 20 mV over a frequency range from 10^5 Hz to 10^{-2} Hz. After each measurement, the samples were kept in sodium chloride solution outside the corrosion cell. The impedance behavior of the samples was simulated using the same software. Polarization measurements were conducted by polarizing the working electrode from an initial potential of -250 mV up to a final potential of +250 mV as a function of the open circuit potential using a scan rate of 1.66 mV/s. To study the reproducibility, each set of experiments was repeated thrice on newly coated samples.

9.3 Results and Discussion

Figure 9.1 and Table 9.1 show the TGA thermograms and thermal

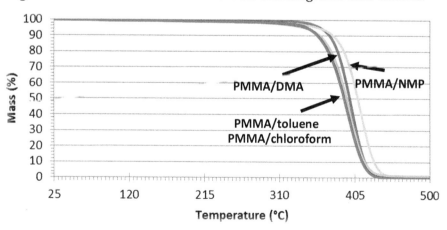

Figure 9.1 TGA thermograms of neat PMMA films developed using different solvents.

Table 9.1 Thermal analysis of PMMA films developed using different solvents

Sample ID	T_{onset} (°C)	Onset y (%)	T_{end} (°C)	End y (%)
PMMA/DMA	377	97.992	417	0.622
PMMA/NMP	386	97.820	429	0.737
PMMA/chloroform	369	97.664	417	1.237
PMMA/toluene	366	97.486	414	0.020

data of the PMMA films prepared using different solvents. The onset of thermal degradation (T_{onset}) temperature was almost similar for PMMA/chloroform and PMMA/toluene. The interaction of DMA and NMP solvents with PMMA increased T_{onset} to 377 °C and 386 °C respectively due to a small extent of residual solvent in the films.

The DSC thermograms of the PMMA films developed from different solvents had almost similar values of T_g, as shown in Figure 9.2.

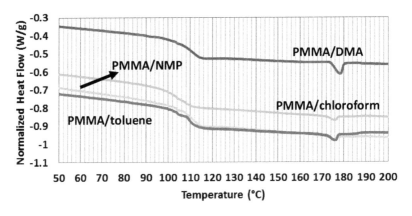

Figure 9.2 DSC thermograms of the PMMA films developed from different solvents.

The presence of a small amount of residual solvent in the range of 2-3% has been reported to have no significant impact on the T_g of the composite films [36]. Therefore, the TGA and DSC analysis showed that the heating cycle was successful in removing the low boiling chloroform and toluene solvents from the PMMA films, but left the high boiling DMA and NMP solvents in a range of less than 2-3%.

FTIR was used to evaluate further the solvent retention in the produced PMMA films. Both DMA and NMP are chemically compatible with PMMA which would also suggest the presence of polar-polar interactions. Figure 9.3 depicts the FTIR spectra of pure PMMA, PMMA precipitated from NMP and PMMA precipitated from DMA. The spectra of PMMA precipitated from solvents indicated the formation of a shoulder peak in the range 1690-1640 cm^{-1}, which corresponded to

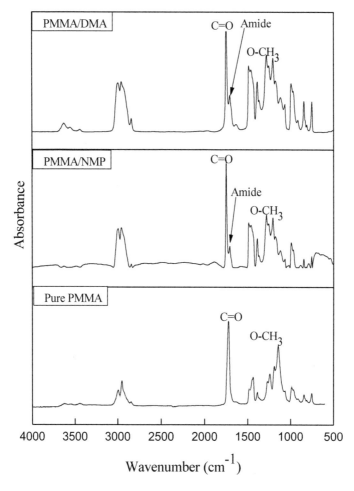

Figure 9.3 FTIR spectra of PMMA and PMMA precipitated from solvents.

the amide linkage found in both solvents. This confirmed the presence of residual solvent in the dry films. Though PMMA dissolved in all solvents, the nanofillers dictated the use of only DMA and NMP due to their adequate dispersion in these solvents. The nanofiller dispersions in toluene and chloroform settled after a few hours, whereas DMA and NMP dispersions were stable for more than two days (Figure 9.4). Therefore, DMA and NMP solvents were used for the nanofiller dispersed composite coating formulations. The bath sonicated rGO samples were not stable for more than one day, but probe sonication stabilized rGO in both DMA and NMP solvents. The presence of hydrophilic functional groups stabilized and dispersed GO adequately in both solvents on application of bath sonication.

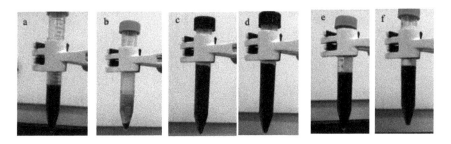

Figure 9.4 rGO immediately after probe sonication in toluene (a), DMA (c) and NMP (e), and after one day probe sonication in toluene (b), DMA (d) and NMP(f).

FTIR spectra in Figure 9.5 show the effect of GO addition in the PMMA matrix. The broad peak present in the GO spectrum at 3400 cm^{-1} could be attributed to the absorbed water and hydroxyl functional groups of GO. Other peaks corresponding to the oxygen functionalities such as carboxylic acid groups, carbonyl groups (~1735 cm^{-1}) and epoxy groups (~1394-1042 cm^{-1}) were also present [37]. The FTIR spectra of PMMA/DMA/5%GO confirmed the effective incorporation of GO in the PMMA host matrix. The C=C peak around 1625 cm^{-1} corresponding to the sp^2 character of the aromatic domains of GO provides convenient spectroscopic evidence of the presence of GO sheets in the PMMA matrix [37]. However, no new peaks corresponding to the formation of bonds between GO and PMMA were detected. Such polar-polar interactions and absence of chemical bonding between GO and PMMA were expected due to the physical mixing of the two during the composite formation. However, the FTIR spectrum of the PMMA/rGO composite was inconclusive and did not yield information about the composite. The FTIR spectrum of rGO reflected the absence of oxygen functionalities which resembled pristine graphene to a large extent. The absence of bands in the region

1300-1100 cm^{-1} suggested that no carboxylic acid or ester groups were present in the structure. Moreover, the broad characteristic peak of the hydroxyl group was absent from the spectrum with only alkene bonds detected around 1618 cm^{-1} [38]. The spectra of both rGO and PMMA/DMA/5%rGO were observed to be almost identical. In this case, as a result of the chemical structure of PMMA and rGO, along with the solvent blending method used to prepare the composite, one can exclude any indication of a chemical interaction between the two. Such behavior is expected due to the lack of enough oxygen functionalities in rGO to interact with the polar bonds of PMMA.

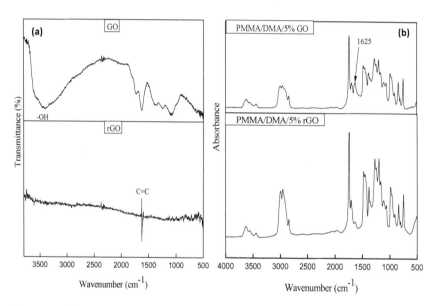

Figure 9.5 FTIR spectra of (a) pure GO and rGO, and (b) PMMA/DMA/5%GO and PMMA/DMA/5%rGO.

The main effect of the two probe sonication durations used for the preparation of PMMA/rGO composites was manifested in the T_{onset}, as seen in Table 9.2. For 25 min sonicated PMMA/DMA/rGO samples, a slight change in T_{onset} was observed only at 5 wt% rGO loading. A continuous decrease in T_{onset} was observed for 60 min sonicated samples with increase in rGO loading. The longer probe sonication time dispersed the rGO sheets to a higher extent in the polymer matrix. The well dispersed rGO sheets in PMMA would create a barrier effect that hinders the diffusion of the gaseous degradation products from the polymer matrix. This effect was expected to increase T_{onset} and grant higher thermal stability to the composites. However, the heat transfer aided by the high thermal conductivity of rGO had an opposite effect on T_{onset}. PMMA/NMP/rGO samples prepared by 25 min probe sonication did not exhibit change in T_{onset} from 386 °C of pure PMMA/NMP. The dispersed rGO interacted strongly with PMMA and NMP via its residual oxygen moieties which prevented the elimination of gaseous degradation products before the T_{onset}.

Table 9.2 Thermal analysis of the PMMA/DMA/rGO samples prepared using different probe sonication durations

Sample ID	T_{onset} (°C)	Onset y (%)	T_{end} (°C)	End y (%)
PMMA/DMA	377	97.992	417	0.622
PMMA/DMA/1%rGO/25	377	91.489	419	0.810
PMMA/DMA/3%rGO/25	377	92.435	418	0.825
PMMA/DMA/5%rGO/25	379	91.241	423	0.790
PMMA/DMA/1%rGO/60	375	98.113	418	0.472
PMMA/DMA/3%rGO/60	371	98.123	412	2.584
PMMA/DMA/5%rGO/60	370	97.741	412	3.825

DSC analysis was conducted to analyze the glass transition temperature of the synthesized PMMA/nanofiller films. As polymer chains are heated to a temperature above their T_g, their segmental mobility becomes easier due to reduced interactions with other neighboring polymer chains. Thus, the higher the T_g of the composite, the more restricted the chain segmental mobility is, which leads to a reduction in the free volume between the polymer chains. A reduction in the free volume/voids between the polymer chains enhances the barrier effect, which positively impacts the anti-corrosion performance of the films. For PMMA/DMA/rGO films, slight changes in T_g were noticed for 1% rGO loading and longer probe sonication time, where the T_g was observed to increase to 111 °C as compared to 107 °C for the composite with 25 min sonication time (Figure 9.6). The samples with higher loading (i.e., 3 and 5% rGO) did not exhibit any change in T_g as the probe sonication time was varied from 25 min to 60 min.

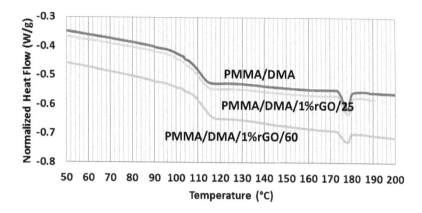

Figure 9.6 DSC thermograms of PMMA/DMA/1%rGO at different probe sonication durations.

However, significant changes in T_g were observed for PMMA/NMP/rGO samples prepared using 25 min probe sonication time (Figure 9.7). The effect was specifically prominent in the case of PMMA/NMP/5%rGO, where the T_g dropped to 96 °C from 109 °C for PMMA/NMP. The observed effect might be a consequence of an enhanced barrier effect imparted by the rGO sheets in the PMMA matrix.

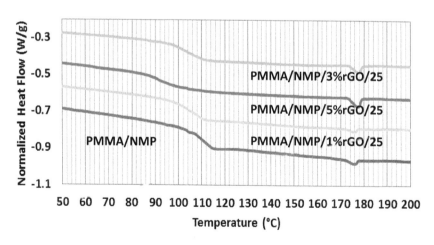

Figure 9.7 DSC thermograms of PMMA/NMP/rGO/25 samples at different rGO loadings.

The rGO sheets increased the tortuous path for the evaporating solvent molecules, thus, blocking the out-diffusion during the drying stage. As a result, the composites retained a certain extent of solvent molecules, which negatively impacted the T_g. The presence of large amounts of residual solvent in the films decreased T_g even below than that of the neat polymer. On the other hand, the longer probe sonication time (60 min) probably resulted in much superior dispersion of rGO in the PMMA matrix. Thus, the T_g values were observed to be similar to the neat polymer owing to the competing factors of restricted chain mobility enhancing the T_g and solvent retention decreasing the T_g. Based on the observations from rGO based films, PMMA/GO films were prepared with DMA solvent. The T_g of the PMMA films was also observed to be unaffected by the addition of GO.

XRD patterns were used to gain insights about the dispersion of the nanofillers in the PMMA matrix [39]. The effect of nanofillers on the long range atomic order of PMMA was also assessed. The XRD patterns of GO and rGO are shown in Figure 9.8a. GO displayed a sharp

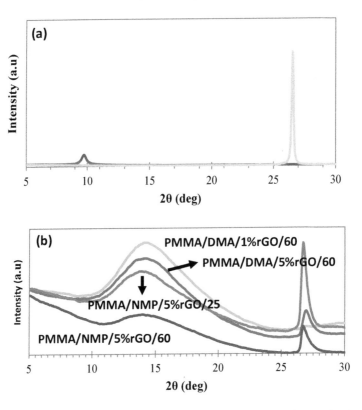

Figure 9.8 XRD patterns of (a) GO and rGO, and (b) PMMA/DMA/rGO composites.

peak at 2θ≈10°, whereas the peak of rGO appeared at 2θ≈26.5°, reflecting the presence of a layered structure in both fillers. This indicated greater interlayer spacing in GO than rGO, attributed to the intercalating oxygen-containing functionalities present in the GO sheets, especially on the edges [39,40]. At 5% loading, the crystalline structure of rGO was retained in the composites and a sharp peak appeared at 2θ≈26.5° (Figure 9.8b). The behavior was observed in all PMMA/5%rGO composites regardless of the solvent and sonication time. This may be attributed to the relatively weak interactions between PMMA and rGO as compared to GO as a result of the non-polar nature of rGO. Figure 9.9 shows the XRD patterns of PMMA and PMMA/GO composites. PMMA displayed characteristic amorphous

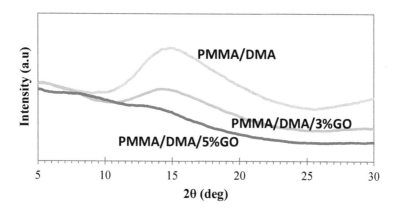

Figure 9.9 XRD patterns of PMMA/DMA and PMMA/DMA/GO samples with different GO fractions.

halo without any sharp peaks corresponding to its non-crystalline structure. The addition of GO did not alter the non-crystalline nature of PMMA even at higher filler loading. Further, the diffraction peak of GO (2θ≈10°) did not appear in the diffractograms, suggesting that GO was exfoliated and dispersed in the PMMA matrix [37,41]. The filler dispersion would have also been promoted by the polar-polar interactions between the oxygen-containing functionalities of GO and PMMA matrix [42,43].

PMMA nanocomposite coatings prepared using different solvents were analyzed for surface conductivity and adhesion strength. Thickness of the synthesized coatings was measured to be in 30-50 μm range. Tape adhesion test, as per ASTM-D3359 standard, was performed to qualitatively evaluate the adhesion of the coatings to the steel substrate (Figure 9.10). It was observed that the coatings prepared using DMA and NMP had 5B and 4B adhesion quality. PMMA coatings prepared using chloroform and toluene were found to be of 3B and 1B adhesion quality, respectively. The adhesion strength depends on the functional groups present in the coating formulation and their interaction with the underlying steel substrate. It has been reported that the polymers containing polar oxygen functionalities, such as the carbonyl group in PMMA, bond with steel via Fe-O=C bonds [44]. Additionally, the high temperature treatment at 270 °C enhances the adhesion by promoting interactions between the polymer functional groups and steel substrate through Fe-C and Fe-O bonds [44]. The electrical conductivity of neat PMMA was measured to be 10^{-14} S/cm (Table 9.3). Addition of rGO to the PMMA matrix did not lead to any noticeable enhancement at 1% and 3% loadings. However at 5% rGO content, the surface resistivity was sharply reduced by 7 orders of magnitude. This result is somewhat similar to the findings of Tripathi *et al.* [45], where the formation of a conducting percolation network throughout the PMMA matrix was reported. Thus, the observed result could be correlated to the quality of filler dispersion, which is dictated by the aforementioned factors. The conductivity of PMMA/GO composites increased only slightly even after 5% loading, as compared to the neat polymer. It was attributed to the presence of the oxygen functionalities (defects) which disrupt the sp^2 arrangement of the graphitic sheets, similar to other research studies reported on these systems [46].

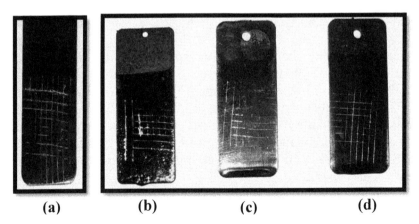

Figure 9.10 Adhesion test images of (a) PMMA/DMA, (b) PMMA/DMA/5%rGO, (c) PMMA/DMA/1%rGO and (d) PMMA/DMA/5%GO.

PMMA Nanocomposite Coatings

Table 9.3 Electrical conductivity of PMMA/DMA/nanofiller composites

Sample	Surface Resistivity (Ω/sq)	Surface Conductivity (S cm^{-1})
Neat PMMA	1.10×10^{11}	9.09×10^{-13}
PMMA/DMA/1%rGO	2.30×10^{11}	4.35×10^{-13}
PMMA/DMA/3%rGO	2.43×10^{11}	4.12×10^{-13}
PMMA/DMA/5%rGO	1.81×10^{4}	5.55×10^{-6}
PMMA/DMA/1%GO	2.90×10^{11}	3.45×10^{-13}
PMMA/DMA/3%GO	3.70×10^{11}	2.70×10^{-13}
PMMA/DMA/5%GO	3.74×10^{11}	2.67×10^{-13}

EIS analysis of the PMMA nanocomposite coatings prepared using NMP and DMA was conducted in 3.5 wt% sodium chloride solution to evaluate the effect of residual solvent, filler loading and probe sonication time on the anti-corrosion performance. Bode plots of PMMA prepared using DMA and NMP solvents, shown in Figure 9.11, exhibited a similar frequency impedance (log Z) of 8.3 Ω cm². After 24 h immersion duration, it was observed to shift respectively to 6.9 Ω cm²

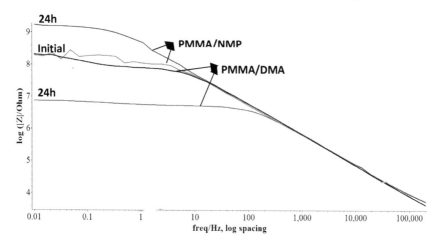

Figure 9.11 Bode plots of PMMA/DMA and PMMA/NMP coatings at different time of immersion in 3.5% NaCl solution.

and 9.0 Ω cm². Low frequency log Z decreased for PMMA/DMA sample exactly like a barrier coating, whereas it increased for PMMA/NMP sample due to the wash out of residual NMP solvent upon exposure to electrolyte. Linear polarization tests given in Figure

9.12 showed that the corrosion current I_{corr} was lower for the PMMA/NMP sample (7.82 10^{-4} µA) than the PMMA/DMA sample (2.04 µA) after 24 h immersion. A more positive E_{corr} value (-163 mV) was exhibited by the NMP based sample compared to the DMA based specimen (-543 mV). These preliminary analysis indicated that even though a small portion of residual NMP existed in the coating, PMMA/NMP coating did not deteriorate as fast as PMMA/DMA based coating. Therefore, PMMA/NMP/5%rGO was chosen to study the effect of probe sonication time on the nanofiller dispersion in the PMMA matrix and corresponding anti-corrosion protective nature.

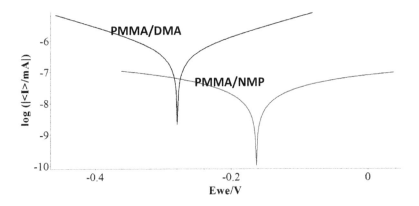

Figure 9.12 Tafel plots of PMMA/DMA and PMMA/NMP coatings.

Bode plots of PMMA/NMP/5%rGO samples prepared with 25 min and 60 min probe sonication durations, shown in Figure 9.13, showed an initial low frequency impedance (log Z) of 4.0 Ω cm² and 9.0 Ω cm² respectively. Though the performance of PMMA/NMP/5%rGO coatings prepared using longer sonication time was better than its counterpart, longer immersion in electrolyte decreased the low frequency log Z to 6.6 Ω cm² which was lower than the pure PMMA/NMP coating. This poor anti-corrosion performance could be attributed to the presence of residual NMP, as the presence of rGO in the PMMA matrix inhibited the complete evaporation of NMP from the coating, as observed in the DSC and TGA analysis.

Figure 9.14 shows the Bode plots of PMMA/DMA/rGO coatings containing 1% and 5% rGO prepared using 60 min probe sonication time. Low frequency log Z of PMMA/DMA/1%rGO coating did not deviate much from the initial value of 8.3 Ω cm² after 24 h immersion in the sodium chloride solution. An extension of the immersion time to

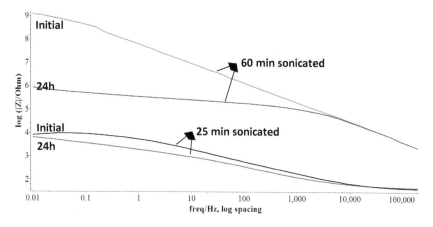

Figure 9.13 Bode plots of PMMA/NMP/5%rGO in 3.5% NaCl solution.

144 h lowered the low frequency log Z to 5.0 Ω cm². The Bode plot of PMMA/DMA/5%rGO/60 sample showed low frequency log Z of 10.5 Ω cm² at the beginning of immersion, which was significantly high compared to the PMMA/DMA/1%rGO/60 coating. The value was observed to remain as such over the course of 24 h analysis. Further immersion for 144 h declined the low frequency modulus to 7.3 Ω cm² which was two orders of magnitude higher than that of 1% rGO coating, indicating enhanced corrosion resistance. The observed protection resulted from the barrier properties imparted by the dispersed hydrophobic rGO nano-sheets in the PMMA matrix. In order to

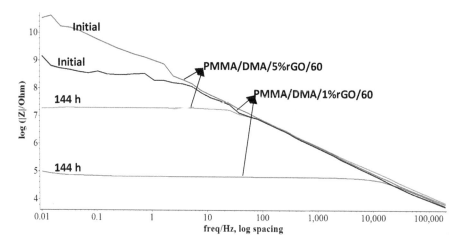

Figure 9.14 Bode plots of PMMA/DMA/rGO/60 in 3.5% NaCl solution.

evaluate the anti-corrosion performance and investigate the processes occurring at the metal-coating interface, the Bode plots of PMMA/DMA/5%rGO/60 were fitted to an equivalent circuit shown in Figure 9.15, where R_s represents the solution resistance or uncompensated resistance, C_{dl} is the double layer capacitance which resembles dipole interactions present on the metal-coating interface [47] and R_{ct} is the polarization resistance caused by the resistance to the

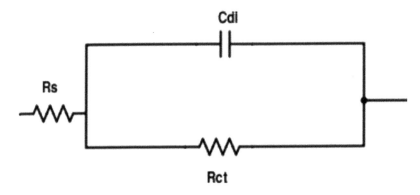

Figure 9.15 Equivalent circuit used to fit the Bode plots of PMMA/DMA/5%rGO/60 coating.

charge transfer in redox reactions that take place on the metal-coating interface. The parameters of the equivalent circuit model are presented in Table 9.4. Noticeably, the C_{dl} value remained almost constant (10^{-9} Ω^{-1} cm^{-2}) throughout the analysis period. This stable double layer capacitance value indicated that no coating degradation or

Table 9.4 Fitted parameters for PMMA/DMA/5%rGO/60 composite coating

Parameters	Initial	6 h	12 h	18 h
log ($|Z|$) (Ω cm^2)	10.51	10.51	10.51	10.51
C_{dl} (Ω^{-1} c)	0.30×10^{-9}	0.29×10^{-9}	0.27×10^{-9}	0.26×10^{-9}
R_{ct} (Ω cm^2)	1.89×10^{27}	94.24×10^9	7.29×10^9	91.16×10^9

Parameters	24 h	72 h	120 h	144 h
log ($|Z|$) (Ω cm^2)	10.51	7.91	7.10	7.31
C_{dl} (Ω^{-1} cm^{-2})	0.28×10^{-9}	0.32×10^{-9}	0.34×10^{-9}	0.41×10^{-9}
R_{ct} (Ω cm^2)	92.19×10^9	59.48×10^6	10.17×10^6	19.44×10^6

underlying corrosion took place. Additionally, C_{dl} is associated with the delamination of the coating, however, C_{dl} did not increase with immersion, which indicated that the electrolyte did not reach the metal surface. Despite the drop in the R_{ct} value after 24 h of immersion, the value remained fairly high, and no visible corrosion products appeared over the course of the analysis period. The parameters of the equivalent circuit remained constant with a high R_{ct} relative to bare steel and a very low C_{dl} value. Overall, these parameters suggested that no electrolyte or aggressive ions reached the metal surface which could be attributed to the barrier properties imparted by the dispersed hydrophobic rGO sheets. The dispersion of the rGO sheets established a tortuous path greatly reducing the accessibility to the metal surface.

The performance of rGO in PMMA coatings was compared by conducting the corrosion analysis on PMMA/DMA/5%GO coating in 3.5 wt% sodium chloride solution. The bode plots, shown in Figure 9.16, showed an initial low frequency log Z of 9.00 Ω cm² which was slightly lower than 5% rGO coating (10.5 Ω cm²). The impedance value dropped to 7.6 Ω cm² at the end of 24 h analysis and to 5.4 Ω cm² after 144 h of immersion. During a similar immersion time PMMA/DMA/5%rGO coatings exhibited a much higher impedance 7.3 Ω cm². It was attributed to the hydrophilic nature of GO which permitted the ingress of electrolyte through the coating. The above analysis proved that the anti-corrosion performance of the 5% rGO coating was superior as compared to the 5% GO coating.

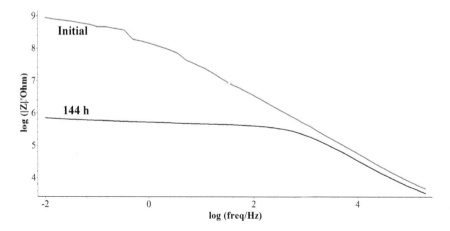

Figure 9.16 Bode plots of PMMA/DMA/5%GO in 3.5% sodium chloride solution.

9.4 Conclusions

The effect of nanofiller on the corrosion protection nature of the polymer coatings was studied by dispersing graphene oxide (GO) and reduced graphene oxide (rGO) in the poly(methyl methacrylate) (PMMA) matrix. The role of solvent and probe sonication time on the filler dispersion as well as corrosion performance was also analyzed. PMMA prepared in N-methylpyrrolidone (NMP) solvent provided better protection than the PMMA/N,N-dimethyl acetamide (DMA) coatings, though residual NMP solvent was observed to exist in the PMMA/NMP coatings. The incorporation of rGO in PMMA/NMP formulation further increased the amount of residual solvent that negatively impacted the glass transition temperature (T_g) and corrosion resistance of the PMMA/NMP/rGO coatings. Such negative effect was not observed in the case of PMMA/DMA/rGO composite coatings. An increased sonication time of 60 min improved the dispersion of rGO in both coating formulations, which, in turn, enhanced the coating performance as well. Among the two nanofillers, rGO imparted superior protection to PMMA coatings than GO. A low frequency impedance of 10.5 Ω cm^2 observed for PMMA/DMA/5%rGO/60 changed to 7.3 Ω cm^2 after 144 h immersion in the 3.5% sodium chloride solution. At the same time, a low frequency impedance of 5.4 Ω cm^2 was observed for PMMA/NMP/5%GO coating. Therefore, it could be suggested that the performance of the PMMA coatings could be enhanced by choosing appropriate solvent and nanofiller. In addition, uniform dispersion of nanofiller in the polymer matrix was also required to obtain nanocomposite coatings with optimal barrier performance.

References

1. Krishnamoorthy, K., Jeyasubramanian, K., Premanathan, M., Subbiah, G., Suk Shin, H., and Jae Kim, S. (2014) Graphene oxide nanopaint. *Carbon*, **72**, 328-337.
2. Prasai, D., Tuberquia, J. C., Harl, R. R., Jennings, G. K., and Bolotin, K. I. (2012) Graphene: Corrosion-inhibiting coating. *ACS Nano*, **6**, 1102-1108.
3. Bohannan, J. (2005) 'Smart coatings' research shows the virtues of superficiality. *Science*, **309**(5733), 376-377.
4. Zhang, Y., Shao, Y., Zhang, T., Meng, G., and Wang, F. (2013) High corrosion protection of a polyaniline/organophilic montmorillonite coating for magnesium alloys. *Progress in Organic Coatings,* **76**, 804-

811.
5. Radhakrishnan, S., Siju, C. R., Mahanta, D., Patil, S., and Madras, G. (2009) Conducting polyaniline–nano-TiO_2 composites for smart corrosion resistant coatings. *Electrochimica Acta*, **54**, 1249-1254.
6. Bagherzadeh, M., and Mousavinejad, T. (2012) Preparation and investigation of anticorrosion properties of the water-based epoxy-clay nanocoating modified by Na^+-MMT and Cloisite 30B. *Progress in Organic Coatings*, **74**, 589-595.
7. Dennis, R. V., Viyannalage, L. T., Gaikwad, A. V., Rout, T. K., and Banerjee, S. (2013) Graphene nanocomposite coatings for protecting low-alloy steels from corrosion. *American Ceramic Society Bulletin*, **92**(5), 18-24.
8. Li, Y., Yang, Z., Qiu, H., Dai, Y., Zheng, Q., Li, J., and Yang, J. (2014) Self-aligned graphene as anticorrosive barrier in waterborne polyurethane composite coatings. *Journal of Materials Chemistry A*, **2**, 14139-14145.
9. Chang, K. C., Hsu, M. H., Lu, H. I., Lai, M. C., Liu, P. J., Hsu, C. H., Ji, W. F., Chuang, T. L., Wei, Y., and Yeh, J. M. (2014) Room-temperature cured hydrophobic epoxy/graphene composites as corrosion inhibitor for cold-rolled steel. *Carbon*, **66**, 144-153.
10. Chang, C. H., Huang, T. C., Peng, C. W., Yeh, T. C., Lu, H. I., Hung, W. I., Weng, C. J., Yang, T. I., and Yeh, J. M. (2012) Novel anticorrosion coatings prepared from polyaniline/graphene composites. *Carbon*, **50**, 5044-5051.
11. Chaudhry, A. U., Mittal, V., and Mishra, B. (2015) Inhibition and promotion of electrochemical reactions by graphene in organic coatings. *RSC Advances*, **5**, 80365-80368.
12. Chang, K., Hsu, C., Lu, H., Ji, W., Chang, C., Li, W., Chuang, T., Yeh, J, Liu, W., and Tsai, M. (2014) Advanced anticorrosive coatings prepared from electroactive polyimide/graphene nanocomposites with synergistic effects of redox catalytic capability and gas barrier properties. *Express Polymer Letters*, **8**(4), 243-255.
13. Liu, S., Gu, L., Zhao, H., Chen, J., and Yu, H. (2016) Corrosion resistance of graphene-reinforced waterborne epoxy coatings. *Journal of Materials Science and Technology*, **32**(5), 425-431.
14. Kirkland, N. T., Schiller, T., Medhekar, N., and Birbilis, N. (2012) Exploring graphene as a corrosion protection barrier. *Corrosion Science*, **56**, 1-4.
15. Zhang, Z., Zhang, W., Li, D., Sun, Y., Wang, Z., Hou, C., Chen, L., Cao, Y., and Liu, Y. (2015) Mechanical and anticorrosive properties of graphene/epoxy resin composites coating prepared by in-situ method. *International Journal of Molecular Sciences*, **16**, 2239-2251.
16. Mo, M., Zhao, W., Chen, Z., Yu, Q., Zeng, Z., Wu, X., and Xue, Q. (2015) Excellent tribological and anti-corrosion performance of polyurethane composite coatings reinforced with functionalized graphene

and graphene oxide nanosheets. *RSC Advances*, **5**, 56486-56497.
17. Sahu, S. C., Samantara, A. K., Seth, M., Parwaiz, S., Singh, B. P., Rath, P. C., and Jena, B. K. (2013) A facile electrochemical approach for development of highly corrosion protective coatings using graphene nanosheets. *Electrochemistry Communications*, **32**, 22-26.
18. Dong, Y., Liu, Q., and Zhou, Q. (2015) Time-dependent protection of ground and polished Cu using graphene film. *Corrosion Science*, **90**, 69-75.
19. Sun, W., Wang, L., Wu, T., Wang, M., Yang, Z., Pan, Y., and Liu, G. (2015) Inhibiting the corrosion-promotion activity of graphene. *Chemistry of Materials*, **27**, 2367-2373.
20. Mišković-Stanković, V., Jevremović, I., Jung, I., and Rhee, K. (2014) Electrochemical study of corrosion behavior of graphene coatings on copper and aluminum in a chloride solution. *Carbon*, **75**, 335-344.
21. Liu, J., Hua, L., Li, S., and Yu, M. (2015) Graphene dip coatings: An effective anticorrosion barrier on aluminum. *Applied Surface Science*, **327**, 241-245.
22. Singh, B. P., Nayak, S., Nanda, K. K., Jena, B. K., Bhattacharjee, S., and Besra, L. (2013) The production of a corrosion resistant graphene reinforced composite coating on copper by electrophoretic deposition. *Carbon*, **61**, 47-56.
23. Ming, H., Wang, J., Zhang, Z., Wang, S., Han, E. H., and Ke, W. (2014) Multilayer graphene: a potential anti-oxidation barrier in simulated primary water. *Journal of Materials Science and Technology*, **30**, 1084-1087.
24. Kumar, S., Sun, L., Caceres, S., Li, B., Wood, W., Perugini, A., Maguire, R., and Zhong, W. (2010) Dynamic synergy of graphitic nanoplatelets and multi-walled carbon nanotubes in polyetherimide nanocomposites. *Nanotechnology*, **21**(10), 105702.
25. Li, J., Cui, J., Yang, J., Li, Y., Qiu, H., and Yang, J. (2016) Reinforcement of graphene and its derivatives on the anticorrosive properties of waterborne polyurethane coatings. *Composites Science and Technology*, **129**, 30-37.
26. Yoo, B. M., Shin, H. J., Yoon, H. W., and Park, H. B. (2014) Graphene and graphene oxide and their uses in barrier polymers. *Journal of Applied Polymer Science*, **131**(1), doi: 10.1002/app.39628.
27. Compton, O. C., Kim, S., Pierre, C., Torkelson, J. M., and Nguyen, S. T. (2010) Crumpled graphene nanosheets as highly effective barrier property enhancers. *Advanced Materials*, **22**(42), 4759-4763.
28. Yang, Y. H., Bolling, L., Priolo, M. A., and Grunlan, J. C. (2013) Super gas barrier and selectivity of graphene oxide-polymer multilayer thin films. *Advanced Materials*, **25**(4), 503-508.
29. Unalan, I. U., Wan, C., Figiel, L., Olsson, R. T., Trabattoni, S., and Farris, S. (2015) Exceptional oxygen barrier performance of pullulan

nanocomposites with ultra-low loading of graphene oxide. *Nanotechnology*, **26**, 275703.

30. Kim, H., Miura, Y., and Macosko, C. W. (2010) Graphene/polyurethane nanocomposites for improved gas barrier and electrical conductivity. *Chemistry of Materials*, **22** (11), 3441-3450.

31. Yousefi, N., Gudarzi, M. M., Zheng, Q. B., Lin, X. Y., Shen, X., Jia, J. J., Sharif, F., and Kim, J. K. (2013) Highly aligned, ultralarge-size reduced graphene oxide/polyurethane nanocomposites: mechanical properties and moisture permeability. *Composites Part A: Applied Science and Manufacturing*, **49**, 42-50.

32. Yucel, O., Unsal, E., Harvey, J., Graham, M., Jones, D. H., and Cakmak, M. (2015) Mechanisms of drying of graphene/poly (amide imide) solutions for films with enhanced gas barrier and mechanical as investigated by real time measurement system. *Journal of Membrane Science*, **495**, 65-71.

33. Bandyopadhyay, P., Park, W. B., Layek, R. K., Uddin, M. E., Kim, N. H., Kim, H. G., and Lee, J. H. (2016) Hexylamine functionalized reduced graphene oxide/polyurethane nanocomposite-coated nylon for enhanced hydrogen gas barrier film. *Journal of Membrane Science*, **500**, 106-114.

34. Yu, B., Wang, X., Xing, W., Yang, H., Song, L., and Hu, Y. (2012) UV-curable functionalized graphene oxide/polyurethane acrylate nanocomposite coatings with enhanced thermal stability and mechanical properties. *Industrial and Engineering Chemistry Research*, **51**, 14629-14636.

35. Wicks, Jr., Z. W., Jones, F. N., Pappas, S. P., and Wicks, D. A. (2007) *Organic Coatings: Science and Technology*, 3rd edition, John Wiley and Sons, USA.

36. Kuila, T., Bose, S., Khanra, P., Kim, N. H., Rhee, K. Y., and Lee, J. H. (2011) Characterization and properties of in situ emulsion polymerized poly (methyl methacrylate)/graphene nanocomposites. *Composites Part A: Applied Science and Manufacturing*, **42**, 1856-1861.

37. Kim, T., Song, B., Cibin, G., Dent, A., Li, L., and Korsunsky, A. M. (2014) A Comparative Spectroscopic Study of Graphene-coated vs Pristine Li(Mn,Ni,Co) Oxide Materials for Lithium-ion Battery Cathodes. *Proceedings of the International MultiConference of Engineers and Computer Scientists*, Hong Kong.

38. Shobhana, E. (2012) X-ray diffraction and UV-visible studies of PMMA thin films. *International Journal of Modern Engineering Research*, **2**(3), 1092-1095.

39. Yang, Y., He, C. E., Tang, W., Tsui, C. P., Shi, D., Sun, Z., Jiang, T., and Xie, X. (2014) Judicious selection of bifunctional molecules to chemically modify graphene for improving nanomechanical and thermal properties of polymer composites. *Journal of Materials Chemistry*

A, **2**, 20038-20047.
40. Dreyer, D. R., Park, S., Bielawski, C. W., and Ruoff, R. S. (2010) The chemistry of graphene oxide. *Chemical Society Reviews*, **39**, 228-240.
41. *In-situ Synthesis of Polymer Nanocomposites*, Mittal, V. (ed.), Wiley VCH, Germany (2012).
42. Gudarzi, M. M., and Sharif, F. (2011) Self assembly of graphene oxide at the liquid-liquid interface: A new route to fabrication of graphene based composites. *Soft Matter*, **7**, 3432-3440.
43. Harb, S. V., Cerrutti, B. M., Pulcinelli, S. H., Santilli, C. V., and Hammer, P. (2015) Siloxane–PMMA hybrid anti-corrosion coatings reinforced by lignin. *Surface and Coatings Technology*, **275**, 9-16.
44. Vakili, H., Ramezanzadeh, B., and Amini, R. (2015) The corrosion performance and adhesion properties of the epoxy coating applied on the steel substrates treated by cerium-based conversion coatings. *Corrosion Science*, **94**, 466-475.
45. Tripathi, S. N., Saini, P., Gupta, D., and Choudhary, V. (2013) Electrical and mechanical properties of PMMA/reduced graphene oxide nanocomposites prepared via in situ polymerization. *Journal of Materials Science*, **48**, 6223-6232.
46. Khanam, P. N., Ponnamma, D., AL-Madeed, M. A. (2015) Electrical properties of graphene polymer nanocomposites. In: *Graphene-Based Polymer Nanocomposites in Electronics*, Sadasivuni, K., Ponnamma, D., Kim, J., and Thomas, S. (eds), Springer, Germany, doi: 10.1007/978-3-319-13875-6_2.
47. *Basics of Electrochemical Impedance Spectroscopy*, Gamry. Online: http://www.princetonappliedresearch.com/download.asbx?AttributeFileId=8406b254-b6d4-4341-9f48-87a04ce7ee3d [accessed 16th March 2019].

10

Anti-microbial Polymeric Materials and Coatings

10.1 Introduction

Controlling the harmful effects of micro-organisms is an important task for improving human health. Major health problems can occur due to the uncontrolled growth of microbes. [1-3]. Most of the polymers and conventional materials used in day to day life generally exhibit no response towards the growth of micro-organisms. Some of them can even act as a medium for the accumulation and propagation of microbes. The surrounding environment also helps in the proliferations and multiplication of these micro-organisms. In this respect, development of anti-microbial materials is very critical [4-6]. Anti-microbial agents can be broadly defined as materials that are capable of neutralizing micro-organisms and preventing their further growth. Low molecular anti-microbial agents are widely used in sterilization of water, anti-microbial drugs, food preservation and soil sterilization [7-16]. The polymeric materials with anti-microbial properties are generally termed as polymeric biocides. The polymeric biocides can be present in many forms like polymeric biocide incorporated fibers or directly extruded as fibers. These fibers can be used in sterile bandages and clothing materials [17-20].

Another approach to inhibit the growth of micro-organisms is the development of anti-microbial surfaces and coatings. These coatings have two-fold action which includes the repulsion of microbes on the surface and neutralizing the microbes in contact with the surface [21-24]. The action is generally achieved by using biocides such as triclosan active chlorine, antibiotics, anti-microbial ammonium, silver compounds, etc. These materials work efficiently in eliminating the microbes, however, can exhaust after continuous usage. This can be prevented by creating surfaces that can produce the biocides catalytically. The catalytic activity can be induced through a number of external stimuli like heat, electricity, optical energy, etc. Another way of preventing the exhaustion of biocides is the use of materials that become active only during a microbial contact. These are called contact-

Liyamol Jacob and Vikas Mittal, Khalifa University of Science and Technology, Abu Dhabi, UAE
© *2021 Central West Publishing, Australia*

active anti-microbial surfaces, and are usually developed by using the anti-microbial polymers tethered (physically or chemically) onto the surface [25-28].

10.2 Different Anti-microbial Polymeric Materials and Coatings

Timofeeva and Kleshcheva [29] recently reviewed the trends in polymeric biocides. Polymeric biocides have been synthesized as a large class of copolymers, either quaternized or functionalized with bioactive groups, anti-microbial macromolecular systems and inherent biocidal polymers [30-36]. In one such study, cycloamine monomer, 3-allyl-5,5-dimethylhydantoin (ADMH), was grafted onto high performing fiber surfaces using a continuous "pad-dry-cure" technique [37]. ADMH readily grafted onto the fibers in the presence of polymers like poly(ethylene glycol)-diacrylate (PEG-DIA). The hydantoin structures of the grafted polymer exhibited regeneratable and long lasting anti-bacterial properties on exposure to chlorine. The polymer was effective against both gram positive and gram negative bacteria. Nigamatullin *et al.* [38] fabricated novel polymer biocides using clay and polymer nanotechnology. The anti-bacterial activity of the cationic surfactant modified organoclay was studied for Staphylococcus aureus (gram positive) and Escherichia coli (gram negative) bacteria. The effective interaction of the organoclay with the cell surface was confirmed to be the reason behind the observed anti-microbial activity.

In another study, amination was used to introduce functional groups on polyacrylonitrile (PAN) [39]. The aminated polymer was further treated with benzaldehyde and its derivatives to evolve anti-microbial properties. The benzaldehyde derivatives used for immobilization on the aminated polymer included 2,4-dihydroxybenzaldehyde and 4-hydroxybenzaldehyde. Staphylococcus aureus, Pseudomonas aeruginosa, Escherichia coli, Salmonella typhi, Cryptococcus neoformans, Candida albicans, Aspergillus niger and Aspergillus flavus were used to study the anti-microbial activity using viable cell counting and cut plug method. It was concluded that the aminated polymer had a durable efficiency in eliminating bacteria and fungi. An increase in the number of phenolic hydroxyl groups of the bioactive moiety was also observed to enhance the biocidal efficiency of the material.

Fluorescence depolarization method was used to study the interaction of poly(hexamethylene biguanide hydrochloride) (PHMB) with phospholipid bilayers [40]. PHMB is a polymer biocide containing biguanide groups in its main chain. The neutral phosphatidylcholine (PC) remained unaffected by the interaction, while the negatively charged bilayers, made by phosphatidylglycerol (PG) alone or mixture of PC of PG, were detected. In the gel phase, the addition of PHMB reduced the fluorescence polarization of diphenylhexatriene embedded in the negatively charged bilayers to a large extent. Polytriazoles (PTAs) based on soyabean oil were also prepared using click polymerization method by Gholami et al. [41]. Two different propyn-functional urethane monomers were reacted with azidated soybean oil (ASBO). To control the biological activity, Cloisite 30B (MMT-N3) and azidated derivatives of a quaternary pyridinium salt (APS) were also grafted on PTAs. PTAs showed a good cytocompatibility towards fibroblast cells in the range of 86 to 98 percentage viability. C. albicans, P. aeruginosa and S. aureus were used to study the anti-bacterial and anti-fungal properties of PTAs. The mutual usage of the modifier materials in coatings showed a high biocidal activity of up to 99% reduction.

Photodynamic therapy (PDT) was used to estimate the anti-bacterial activity of novel oligo(thiophene ethynylene) (OTE) against Escherichia coli, Ralstonia solanacearum, Staphylococcus epidermidis and Staphylococcus aureus in vitro [42]. OTE showed remarkably broad spectrum of high anti-bacterial activity after white light irradiation. 52, 24, 13 and 8 ng/mL half inhibitory concentration (IC50) values were obtained for R. solanacearum, E. coli, S. epidermidis and S. aureus respectively after 30 min white light irradiation. The mechanism of action of this material is depicted in Figure 10.1. Particularly, at a concentration of 180 ng/mL, OTE exhibited a strong and specific dark killing capability against S. aureus for 30 min.

In another report among several other novel studies on polymeric biocides [43-50], a bacterial cellulose based material was combined with the polymeric biocide polyhexamethylene guanidine hydrochloride (PHMG-Cl) into a porous structure [50]. The bacterial cellulose exhibited effective saturation in the polymer scaffold. The modified PHMG-Cl biocide bacterial cellulose film specimens exhibited effective efficiency against a range of micro-organisms like yeast Pseudomonas syringae PV. Tomato DC 3000, phytopathogenic Xanthomonas campestris PV. Campestris IMBG 299, Klebsiella pneumonia IMBG 233 and Staphylococcus aureus. Further, the material prevented biofilm formation by prokaryotic and eukaryotic organisms. The biocidal activity of the composite depended on the rate of biocide release.

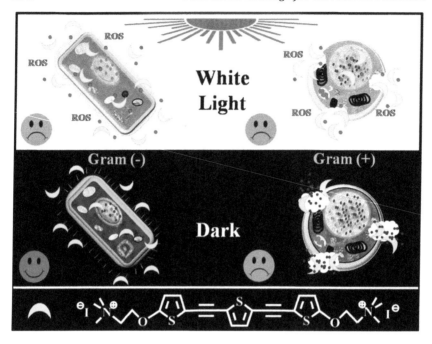

Figure 10.1 Mechanism for OTE cytotoxicity under white light and in dark. Reproduced from Reference 42 with permission from American Chemical Society.

Mycobacteria is one of the most prominent micro-organisms causing a variety of diseases and even mortality. It has resistance towards most of the commonly used disinfectants including low molecular quaternary polymeric biocides owing to its distinct cell wall structure. For studying the biocidal activity on mycobacteria, non-quaternary protonated polydiallylamines (PDAAs) were synthesized [51]. The non-quaternary PDAA acted by breaking the inner membrane permeability in M. Smegmatis cells within 20 min of contact. In another study, bromoacetyl chloride or chloroacetyl chloride modified chitosan were synthesized [52]. Triphenyl phosphine and triethyl amine were used to quaternize the haloacetylated chitosan derivatives. The new derivatives exhibited much better efficiency than chitosan and other commonly used derivatives against gram positive and gram negative bacteria and fungi. The death of the cells was achieved by the complete leakage of cytoplasm. Charged metallopolymers have also been studied for the effective lysing of bacterial cell walls and inhibiting the activity of β-lactamase in Methicillin-resistant Staphylococcus aureus (Figure 10.2) [53]. The formation of unique ion pairs between the cationic cobaltocenium moieties and carboxylate anions inhibited β-lactamase hydrolysis in various β-lactam antibiotics, including cefazolin, amoxicillin ampicillin and penicillin-G. Other biocides with an ability to neutralize multidrug resistant bacteria have also been reported [54-57].

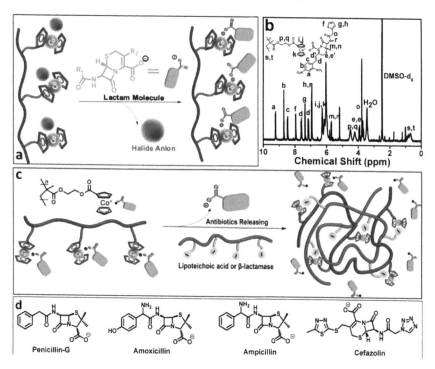

Figure 10.2 (a) Cationic cobaltocenium-containing polymers and β-lactam forming ion-pairs between antibiotics, (b) cationic cobaltocenium-containing polymer and nitrocefin ion-pair's 1 H NMR spectrum, (c) β-lactamases and lipoteichoic acid releasing antibiotic from antibiotic-metallopolymer ion pairs and (d) β-lactam antibiotics utilized in the study. Reproduced from Reference 53 with permission from American Chemical Society.

The formation of biofilms make the destruction of micro-organisms almost 1000 times more difficult. Out of the various methods to prevent the formation of biofilms, the use of coatings capable of releasing anti-microbial agents and containing anti-adhesive components is common. Toxicity of the released anti-microbial agent towards microbes is important in healthcare. Anti-adhesive coatings, on the hand, generally do not interfere with the growth, instead these make it difficult for the biofilm to form. In the body, protein adhesion is the predecessor for any bodily bacterial adhesion, and anti-adhesive coatings are designed to be resistant to such protein adhesion [58-62]. In another study combining the use of two or more operational entities, Gehring *et al.* [63] reported mesoporous organosilicas (PMOs) based on co-condensation of sol-gel precursors with bridging phenyl derivatives $R_{F1,2}C_6H_3[Si(O^{iso}Pr)_3]_2$ (Figure 10.3). Subsequently, PMOs containing high density of thiol and sulfonic acid units were prepared as mesoporous nanoparticles. Another study has also reported the use of triazole/nanotubes conjugates as filler for gel-coat nanocomposites [64].

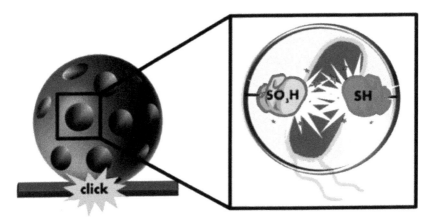

Figure 10.3 Schematic representation of the superacid and click functionalities. Reproduced from Reference 63 with permission from American Chemical Society.

Finlay *et al.* [65] reported biofilms of Navicula incerta prepared from a novel culture system using an open channel flow with adjustable bed shear stress values (0-2.4 Pa). Polydimethylsiloxane elastomer (PDMSe) and glass were used to study and differentiate the biofilm development. The growth of biofilm was prevented by a critical shear stress of 1.3-1.4 Pa on glass. In the case of elastomer, even at 2.4 Pa, the biofilm continued to grow. Schultz *et al.* [66] also presented skin-friction results for fouling-release (FR) hull coatings and suggested a new effective roughness length scale (keff) for biofilms. Ciliate assemblages have also been studied to compare anti-fouling (AF) and FR coatings [67]. In microbial fouling, toxicity of AF coatings is an issue as compared to the FR coatings. The toxic effects shown by AF coatings caused important differences in ciliate species assembly.

In another study, biocidal coatings like controlled depletion polymer (CDP), self-polishing hybrid (SPH) and self-polishing copolymer (SPC) were studied for the microbial community composition of the biofilms developed on them after immersion in sea for a year [68]. Distinct bacterial structures on various coatings were revealed by pyrosequencing of 16S rRNA's genes.

Polymer biocides have considerable importance in textile industry [69]. The need of enhanced hygiene products has paved the way for anti-microbial textiles which have found their use in protective clothing, home furnishing, sportswear and wound-dressings [70-76]. Simultaneous one-step sono-chemical deposition of chitosan and ZnO nanoparticles (NPs) on cotton fabric to achieve anti-microbial textiles was reported by Petkova et al. [77]. The biocompatibility and anti-microbial efficacy were further improved by optimizing the reaction duration and concentration. 2 mM ZnO NPs suspension in a 30 min sono-chemical process imparted the cotton fabric the highest anti-bacterial activity. This ability was further enhanced by adding the same amount of chitosan during the deposition. Figure 10.4 shows the scanning electron microscopy (SEM) images of the cotton fabric with ZnO and chitosan nanoparticles.

Seeded semi-continuous emulsion copolymerization was used for the synthesis of anti-microbial latexes with core shell structure [78]. The latex was based on anti-microbial macromonomer (GPHGH) and hydrophobic acrylate monomers. The polymerization was carried out in the presence of a cationic surfactant. In another study, covalent mechanism was used to synthesize anti-microbial L-cysteine (L-Cys)-functionalized cotton [79]. The covalently bound cotton threads showed superior antimicrobial capacities of 83% and 89% against S. aureus K. pneumoniae respectively.

Inorganic-organic hybrid polymers modified/filled with ZnO nanoparticles were attached to cotton/polyester (65/35%) and cotton (100%) fabrics by Farouk et al. [80]. The composite material showed superior ability to inhibit a wide range of bacterial infections. In another study, peppermint releasing alginate nanocapsules made by micro-emulsion method were applied to cotton fabrics using microwave curing method [81]. Various parameters were optimized to obtain a fabric with optimal anti-microbial properties. In optimum conditions, the fabric showed 100% reduction in the amount of E. coli and S. aureus bacteria. Even after 25 washing cycles, the composite were observed to retain 16% of the initial amount of peppermint oil. Composites containing chitosan (CH) and alginate (ALG) layers were prepared using layer by layer method and attached to cotton fabrics using electrostatic method [82]. The composites had bacteriostatic effects on bacteria, viz. *Klebsiella pneumonia* and Staphylococcus *aureus.* The composite with five multilayers (CH/ALG/CH/ALG/CH) was observed to be most effective for bacteriostatic inhibition.

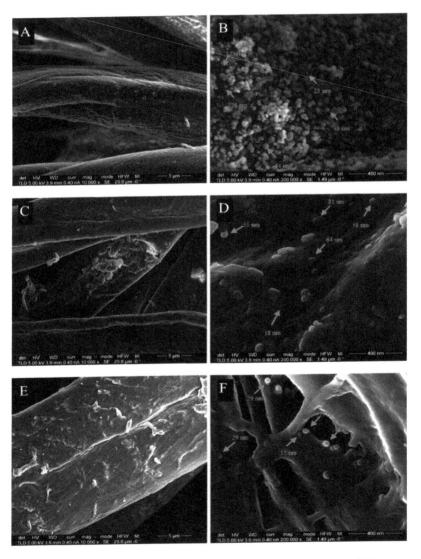

Figure 10.4 SEM images of cotton fabric coated with (A,B) ZnO, (C,D) ZnO/CS and (E,F) CS. Reproduced from Reference 77 with permission from American Chemical Society.

Contact lenses, hip implants, cardiac pacemakers, catheters, etc., have brought a tremendous change in the medical industry. The presence of biofilms in these implants creates issues due to various infections [83,84]. Microbial infection causes almost 20% mortalities worldwide, and 80% of these result from biofilm formation [85,86]. A range of organisms has been specifically implicated in device/biomaterial-related infections [87-90]. It involves both fungal groups and bacteria [91,92]. New bifunctional amphiphilic random copolymers were developed for use in medical devices as coatings [93], by polymerization of an anti-microbial cationic monomer, antioxidant and anti-microbial hydrophobic monomer (containing hydroxytyrosol, HTy). The authors observed that the lowest minimal inhibitory concentration against Staphylococcus epidermidis was shown by the copolymer having the highest HTy molar content. In another study, PVC, PU and silicone were coated with anti-microbial nanoparticles [94]. Specifically, the *in-situ* coating of selenium (Se) nanoparticles rendered these polymers anti-bacterial abilities. The density of the Se atoms on the polymer surface had a direct influence on its anti-microbial activity. Polymer-peptide conjugates were also used for coating biomaterial implants and devices for infection resistance by Muszanska *et al.* (Figure 10.5) [95]. Anti-microbial peptides (AMP) were used to functionalize anti-adhesive polymer brushes made from block copolymer Pluronic F-127 (PF127). Adhesion and spreading of host tissue cells were promoted by arginine-glycine-aspartate (RGD) peptides. Without hampering tissue compatibility, the coatings composed of a suitable ratio of the functional constituents: PF127, PF127 modified with AMP and PF127 modified with RGD showed good anti-adhesive and bactericidal properties.

The insoluble disinfectants initiate the microbial inactivation mechanism by the interaction of the reactive species in the bulk phase. Polymeric disinfectants are an ideal choice for this application. [96-100]. In one such study, thiol functionalized PVDF membranes were developed with surface assembled silver nanoparticles [101]. The formation of the nanocluster assembly of silver nanoparticles was facilitated by the esterification reaction between alkaline treated PVDF membrane (TGA-PVDF) and thioglycolic acid (TGA). In another study, Ahmed *et al.* [102] reported the improved bactericidal behavior of the polyvinyl-N-carbazole (PVK) and of single-walled carbon nanotubes (SWNTs) (97:3 wt% ratio PVK:SWNTs) composite, as compared to 100% SWNTs coated membranes. The PVK-SWNTs membrane was completely non-toxic to fibroblast cells as opposed to pure SWNTs, which showed acute levels of toxicity to the exposed cell lines.

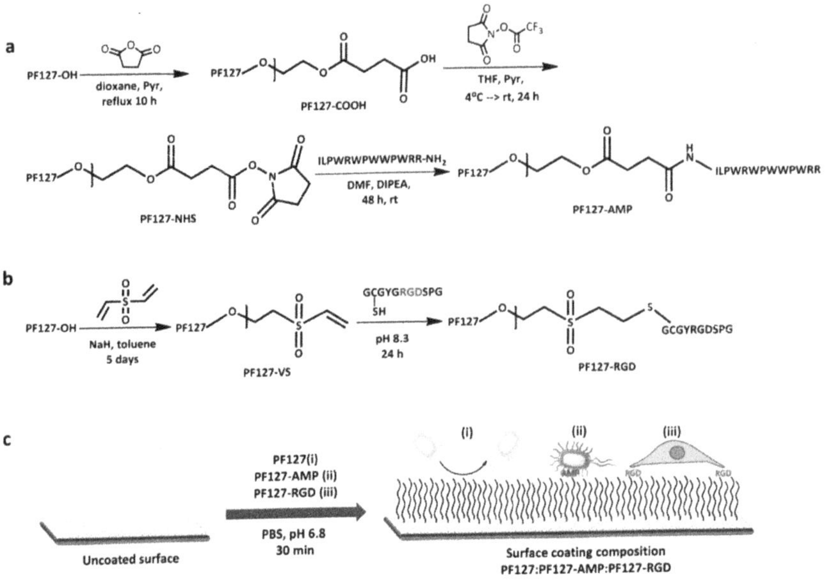

Figure 10.5 (a) PF127-AMP, (b) PF127-RGD and (c) the triple activity coating made by immobilizing the conjugates on a silicone rubber surface using dip-coating. Reproduced from Reference 95 with permission from American Chemical Society.

Bollmann et al. [103] studied polyacrylate-water partitioning of biocidal compounds. The polyacrylate-water partition constants were observed to be predominantly below octanol-water partition constants. Capsaicin-mimic materials were also reported by Zhang et al. [104] as anti-fouling membranes for water treatment. Polyaniline Th(IV) tungstomolybdophosphate (PANI/TWMP) nanocomposite ion exchanger was used as a biocidal membrane by Sharma et al. [105], which exhibited abilities to filter heavy metals from water.

Highly efficient phosphate scavenger to realize nutrient-starvation anti-bacteria was designed using La(OH)$_3$ nanorods immobilized in polyacrylonitrile (PAN) nanofibers (PLNFs). Electrospinning and a subsequent in-situ surfactant-free precipitation method were used to synthesize the composite. The La(OH)$_3$ nanorods with PAN protection showed 8 times more phosphate capture capacity than La(OH)$_3$ nanocrystals without PAN protection. Figure 10.6 illustrates the nutrient-starvation anti-bacteria by PLNFs.

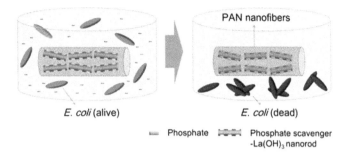

Figure 10.6 Schematic illustration of the nutrient-starvation anti-bacteria by PLNFs. Reproduced from Reference 106 with permission from American Chemical Society.

The idea of anti-microbial packaging has received significant attention in view of its capability to upgrade food safety [107]. Gemili *et al.* [108] incorporated cellulose acetate films with lysosome for obtaining anti-microbial packaging materials. By changing the composition of the initial casting solution, the structure of the films could be changed from highly asymmetric to densely porous, which, in turn, facilitated the controlled release of lysosome. Dutta *et al.* [109] also reviewed various preparative methods and anti-microbial activity of chitosan based films for food applications. In another study, lysosome was incorporated into various polymers, viz. cellulose triacetate (CTA) films, nylon 6,6 pellets, polyvinyl alcohol (PVOH) beads, etc. [110]. Against a suspension of dried *Micrococcus lysodeikticus* cells, polyvinyl alcohol and nylon 6,6 yielded low activity, while CTA yielded the highest activity for dried cells in 30 min. Jin and Zhang [111] reported nisin incorporated biodegradable polylactic acid (PLA) for use in anti-microbial food packaging. Liquid foods (liquid egg white and orange juice) and culture media were used for studying the anti-microbial activity of PLA/nisin films against Salmonella Enteritidis, Escherichia coli O157:H7 and Listeria monocytogenes. The nisin particles inhibited L. monocytogenes growth in the liquid egg white and culture media.

10.3 Conclusions

Modern society is faced with a number of infections, and their control is of high importance. Disinfection and anti-microbial surfaces are the main strategies to eradicate these infections. Development of resistant microbial strains and environmental pollution caused by the disinfectants make them less attractive for everyday use. On the other hand, anti-microbial coatings, which release biocides on coming in contact with microbes, present much effective alternatives. In this respect, the use of polymers and polymeric coatings with biocidal properties is very promising and is envisaged to evolve further in the coming years.

References

1. Chang, C. C., Lin, C. K., Chan, C. C., Hsu, C. S., and Chen, C. Y. (2006) Photocatalytic properties of nanocrystalline TiO_2 thin film with Ag additions. *Thin Solid Films*, **494**, 274-278.
2. Dastjerdi, R., Mojtahedi, M. R. M., and Shoshtari, A. M. (2008) Investigating the effect of various blend ratios of prepared masterbatch containing Ag/TiO2 nanocomposite on the properties of bioactive continuous filament yarns. *Fibers and Polymers*, **9**, 727-734.
3. Yuranova, T., Mosteo, R., Bandara, J., Laubb, D., and Kiwi, J. (2006) Self-cleaning cotton textiles surfaces modified by photoactive SiO_2/TiO_2 coating. *Journal of Molecular Catalysis A: Chemical*, **244**, 160-167.
4. Nersisyan, H. H., Lee, J. H., Son, H. T., Won, C. W., and Maeng, D. Y. (2003) A new and effective chemical reduction method for preparation of nanosized silver powder and colloid dispersion. *Materials Research Bulletin*, **38**, 949-956.
5. Yeo, S. Y., Lee, H. J., and Jeong, S. H. (2003) Preparation of nanocomposite fibers for permanent antibacterial effect. *Journal of Materials Science*, **38**, 2143-2147.
6. Xie, Y., Ye, R., and Liu, H. (2006) Synthesis of silver nanoparticles in reverse micelles stabilized by natural bio surfactant. *Colloids and Surfaces A: Physicochemical Engineering Aspects*, **279**, 175-178.
7. Shahverdi, A. R., Minaeian, S., Shahverdi, H. R., Jamalifar, H., and Nohi, A. A. (2007) Rapid synthesis of silver nanoparticles using culture supernatants of Enterobacteria: a novel biological approach. *Process Biochemistry*, **42**, 919-923.
8. Jeong, S. H., Hwang, Y. H., and Yi, S. C. (2005) Antibacterial properties of padded PP/PE nonwovens incorporating nano-sized silver colloids. *Journal of Materials Science*, **40**, 5413-5418.
9. Wright, T. (2002) Alphasan: A thermally stable silver-based inorganic antimicrobial technology. *Chemical Fiber International*, **52**, 125.
10. Kumar, V. S., Nagaraja, B. M., Shashikala, V., Padmasri, A. H., Madhavendra, S. S., and Raju, B. D. (2004) Highly efficient Ag/C catalyst prepared by electro-chemical deposition method in controlli-

ng microorganisms in water. *Journal of Molecular Catalysis A: Chemical*, **223**, 313-319.
11. Vigo, T. L. (1994) *Textile Processing and Properties*, 11th volume, Elsevier, USA.
12. Yamamoto, T., Uchida, S., Kurihara, Y., and Nakayama, I. (1994) Jpn patent 94-204681.
13. Kenawy, E. R., Abdel-Hay, F. I., El-Shanshoury, A. R., and El-Newehy, M. H. (1998) Biologically active polymers: synthesis and antimicrobial activity of modified glycidyl methacrylate polymers having a quaternary ammonium and phosphonium groups. *Journal of Controlled Release*, **50**, 145-152.
14. Eknoian, M. W., Worley, S. D., and Harris, J. M. (1998) New biocidal N-halamine-PEG polymers. *Journal of Bioactive and Compatible Polymers*, **13**, 136-145.
15. Kanazawa, A., Ikeda, T., and Endo, T. (1993) Polymeric phosphonium salts as a novel class of cationic biocides. III. Immobilization of phosphonium salts by surface photografting and antibacterial activity of the surface-treated polymer films. *Journal of Polymer Science, Part A: Polymer Chemistry*, **31**, 1467-1472.
16. Ilker, M. F., Nuesslein, K., Tew, G. N., and Coughlin, E. B. (2004) Tuning the hemolytic and antibacterial activities of amphiphilic polynorbornene derivatives. *Journal of the American Chemical Society*, **126**, 15870-15875.
17. (a) Gabriel, G. J., Madkour, A. E., Dabkowski, J. M., Nelson, C. F., Nusslein, K., Tew, G. N. (2008) Synthetic mimic of antimicrobial peptide with nonmembrane-disrupting antibacterial properties. *Bio Macromolecules*, **9**, 2980-2983; (b) Lienkamp, K., Madkour, A. E., Musante, A., Nelson, C. F., Nusslein, K., Tew, G. N. (2008) Antimicrobial polymers prepared by ROMP with unprecedented selectivity: a molecular construction kit approach. *Journal of American Chemical Society*, **130**, 9836-9843.
18. Zasloff, M. (2002) Antimicrobial peptides of multicellular organisms. *Nature*, **415**, 389-395.
19. Shai, Y. (1999) Mechanism of the binding, insertion and destabilization of phospholipid bilayer membranes by alpha-helical antimicrobial and cell non-selective membrane-lytic peptides. *Biochimica et Biophysica Acta (BBA) - Biomembranes.* **1462**, 55-70.
20. Mowery, B. P., Lee, S. E., Kissounko, D. A., Epand, R. F., Epand, R. M., Weisblum, B., Stahl, S. S., and Gellman, S. H. (2007) Mimicry of antimicrobial host-defense peptides by random copolymers. *Journal of American Chemical Society*, **129**, 15474-15476.
21. Gelman, M. A., Weisblum, B., Lynn, D. M., and Gellman, S. H. (2004) Biocidal activity of polystyrenes that are cationic by virtue of protonation. *Organic Letters*, **6**, 557-560.
22. Palermo, E. F., and Kuroda, K. (2009) Chemical structure of cationic

groups in amphiphilic polymethacrylates modulates the antimicrobial and hemolytic activities. *Biomacromolecules*, **10**, 1416-1428.
23. Palermo, E. F., Sovadinova, I., and Kuroda, K. (2009) Structural determinants of antimicrobial activity and biocompatibility in membrane-disrupting methacrylamide random copolymers. *Biomacromolecules*, **10**, 3098-3107.
24. Huttinger, K. J., Muller, H., and Bomar, M. T. (1982) Synthesis and effect of carrier-bound disinfectants. *Journal of Colloid Interface Science*, **88**, 274-285.
25. Ren, H., Du, Y., Su, Y., Guo, Y., Zhu, Z., and Dong, A. (2018) A review on recent achievements and current challenges in antibacterial electrospun N-halamines. *Colloid and Interface Science Communications*, **24**, 24-34.
26. Liang, J., Barnes, K., Akdag, A., Worley, S. D., Lee, J., Broughton, R. M., and Huang, T. S. (2007) Improved antimicrobial Siloxane. *Industrial & Engineering Chemistry Research*, **46**, 1861-1866.
27. Charville, G. W., Hetrick, E. M., Geer, C. B., and Schoenfisch, M. H. (2008) Reduced bacterial adhesion to fibrinogen-coated substrates via nitric oxide release. *Biomaterials*, **29**, 4039-4044.
28. Ferrazzano, G. F., Roberto, L., Amato, I., Cantile, T., Sangianantoni, G., and Ingenito, A. (2011) Antimicrobial properties of green tea extract against cariogenic microflora: an in vivo study. *Journal of Medicinal Food*, **14**, 907-911.
29. Timofeeva, L., and Kleshcheva, N. (2011) Antimicrobial polymers: mechanism of action, factors of activity, and applications. *Applied Microbiology and Biotechnology*, **89**, 475-492.
30. Gabriel, G. J., Maegerlein, J. A., Nelson, C. F., Dabkowski, J. M., Eren, T., Nüsslein, K., and Tew, G. N. (2009) Comparison of facially amphiphilic versus segregated monomers in the design of antibacterial copolymers. *Chemistry - A European Journal*, **15**, 433-439.
31. Qi, F., Qian, Y., Shao, N., Zhou, R., Zhang, S., Lu, Z., Zhou, M., Xie, J., Wei, T., Yu, Q., and Liu, R. (2019) Practical preparation of infection-resistant biomedical surfaces from antimicrobial β-peptide polymers. *ACS Applied Materials & Interfaces*, **11**(21), 18907-18913.
32. Gilbert, P., and McBain, A. J. (2003) An evaluation of the potential impact of the increased use of biocides within consumer products upon the prevalence of antibiotic resistance. *Clinical Microbial Reviews*, **16**, 189-208.
33. Gilbert, P., and Moore, L. E. (2005) Cationic antiseptics: diversity of action under a common epithet. *Journal of Applied Microbiology*, **99**, 703-715.
34. Grapski, J. A., and Cooper, S. L. (2001) Synthesis and characterization of non-leaching biocidal polyurethanes. *Biomaterials*, **22**, 2239-2246.
35. Haldar, J., An, D., Alvarez de Cienfuegos, L., Chen, J., and Klibanov, A.

M., (2006) Polymeric coatings that inactivate both influenza virus and pathogenic bacteria. *Proceedings of the National Academy of Sciences of USA*, **103**, 17667-17671.
36. Tew, G. N., Scott, R. W., Klein, M. L., and DeGrado, W. F. (2010) De novo design of antimicrobial foldamers and small molecules: from discovery to practical application. *Accounts of Chemical Research*, **43**, 30-39.
37. Sun, Y., and Sun, G. (2003), Novel refreshable N-halamine polymeric biocides: Grafting hydantoin-containing monomers onto high performance fibers by a continuous process. *Journal of Applied Polymer Science*, **88**, 1032-1039.
38. Nigmatullin, R., Gao, F., and Konovalova, V. J. (2008) Polymer-layered silicate nanocomposites in the design of antimicrobial materials. *Journal of Material Science*, **43**, 5728-5733.
39. Alamri, A., El-Newehy, M. H., and Al-Deyab, S. S. (2012) Biocidal polymers: synthesis and antimicrobial properties of benzaldehyde derivatives immobilized onto amine-terminated polyacrylonitrile. *Chemistry Central Journal*, **6**, 111.
40. Ikeda, T., Tazuke, S., and Watanabe, M. (1983) Interaction of biologically active molecules with phospholipid membranes, *Biochimica ET Biophysica Acta (BBA) - Biomembranes*, **735**, 380-386.
41. Gholami, H., Yeganeh, H., Gharibi, R., Jalilian, M., and Sorayya, M. (2015) Catalyst free-click polymerization: A versatile method for the preparation of soybean oil based poly1,2,3-triazoles as coatings with efficient biocidal activity and excellent cytocompatibility. *Polymer*, **62**, 94-108.
42. Zhao, Q., Li, J., Zhang, X., Li, Z., and Tang, Y. (2016) Cationic Oligo(thiophene ethynylene) with broad-spectrum and high antibacterial efficiency under white light and specific biocidal activity against S. aureus in dark. *ACS Applied Materials and Interfaces*, **8**, 1019-1024.
43. Kawabata, N. (1992) Capture of microorganisms and viruses by pyridinium-type polymers and application to biotechnology and water-purification. *Progress in Polymer Science*, **17**, 1-34.
44. Dizman, B., Elasri, M. O., and Mathias, L. J. (2005) Synthesis, characterization, and antibacterial activities of novel methacrylate polymers containing norfloxacin. *Biomacromolecules*, **6**, 514-520.
45. Lawson, M. K. C., Shoemaker, R., Hoth, K. B., Bowman, C. N., and Anseth, K. S. (2009) Polymerizable vancomycin derivatives for bactericidal biomaterial surface modification: structure-function evaluation. *Biomacromolecules*, **10**, 2221-2234.
46. Waschinski, C. J., and Tiller, J. C., (2005) Poly(oxazoline)s with telechelic antimicrobial functions. *Biomacromolecules*, **6**, 235-243.
47. Turos, E., Shim, J. Y., Wang, Y., Greenhalgh, K., Reddy, G. S. K., Dickey, S., and Lim D. V. (2007) Antibiotic-conjugated polyacrylate nanoparticles: New opportunities for development of anti-MRSA agents.

Bioorganic & Medicinal Chemistry Letters, **17**, 53-56.
48. Tashiro, T. (2001) Antibacterial and bacterium adsorbing macromolecules. *Macromolecular Materials and Engineering*, **286**, 63-87.
49. Nathan, A., Zalipsky, S., Ertel, S. I., Agathos, S. N., Yarmush, M. L., and Kohn, J. (1993) Copolymers of lysine and polyethylene glycol: a new family of functionalized drug carriers. *Bioconjugate Chemistry*, **4**, 54-62.
50. Kukharenko, O., Bardeau, J. F., Zaets, I., Ovcharenko, L., Tarasyuk, O., Porhyn, S., Mischenko, I., Vovk, A., Rogalsky, S., and Kozyrovska, N. (2014) Promising low cost antimicrobial composite material based on bacterial cellulose and poly hexamethylene guanidine hydrochloride. *European Polymer Journal*, **60**, 247-254.
51. Timofeeva, L. M., Kleshcheva, N. A., and Shleeva, M. O. (2015) Non-quaternary poly (diallylammonium) polymers with different amine structure and their biocidal effect on Mycobacterium tuberculosis and Mycobacterium smegmatis. *Applied Microbiological Biotechnology*, **99**, 2557-2571.
52. El-Newehy, M. H., Kenawy, E. R., and Al-Deyab S.S., (2014) Biocidal polymers: preparation and antimicrobial assessment of immobilized onium salts onto modified chitosan. *International Journal of Polymeric Materials and Polymeric Biomaterials*, **63**, 758-766.
53. Zhang, J., Chen, Y. P., Miller, K. P., Ganewatta, M. S., Bam, M., Yan, Y., Nagarkatti, M., Decho, A. W., and Tang, C. (2014) Antimicrobial metallo polymers and their bioconjugates with conventional antibiotics against multidrug-resistant bacteria. *Journal of American Chemical Society*, **136**, 4873-4876.
54. Sellenet, P. H., Allison, B., Applegate, B. M., and Youngblood, J. P. (2007) Blood synergistic activity of hydrophilic modification in antibiotic polymers. *Biomacromolecules*, **8**(1), 19-23.
55. *World Health Statistics 2009*, World Health Organization (2009). Online: https://www.who.int/whosis/whostat/2009/en/ [accessed 19th June 2019].
56. Gabriel, G. J., Som, A., Madkour, A. E., Eren, T., Tew, G. N., (2007) Infectious disease: connecting innate immunity to biocidal polymers. *Materials Science and Engineering R: Reports*, **57**, 28-64.
57. Spielberg, B., Powers, J. H., Brass. E. P., Miller. L. G., and Edwards, Jr., J. E. (2004) Trends in antimicrobial drug development: Implications for the future. *Clinical Infectious Diseases*, **38**, 1279-1286.
58. Farzinfar, E., and Paydayesh, A. (2018) Investigation of polyvinyl alcohol nanocomposite hydrogels containing chitosan nanoparticles as wound dressing. *International Journal of Polymeric Materials and Polymeric Biomaterials*, **68**, 628-638.
59. Evans R. C., and Holmes, C. J. (1987) Effect of vancomycin hydrochloride on Staphylococcus epidermidis biofilm associated with silicone elastomer. *Antimicrobial Agents and Chemotherapy*, **31**, 889-

894.
60. Abel, T., Cohen, J. L. I., Engel, R., Filshtinskaya, M., Melkonian, A., and Melkonian, K. (2002) Preparation and investigation of antibacterial carbohydrate-based surfaces. *Carbohydrate Research*, **337**, 2495-2499.
61. Ewald, A., Glückermann, S. K., Thull, R., and Gbureck, U. (2006) Antimicrobial titanium/silver PVD coatings on titanium. *Biomedical Engineering Online*, **5**, 22.
62. Barasch, A., Elad, S., Altman, A., Damato, K., and Epstein, J. (2006) Antimicrobials, mucosal coating agents, anesthetics, analgesics, and nutritional supplements for alimentary tract mucositis. *Support Care Cancer*, **14**, 528-532.
63. Gehring, J., Schleheck, D., Trepka, B., and Polarz, S. (2015) Mesoporous organosilica nanoparticles containing superacid and click functionalities leading to cooperativity in biocidal coatings. *ACS Applied Materials and Interfaces*, **7**, 1021-1029.
64. Iannazzo, D., Pistone, A., Visco, A., Galtieri, G., Giofrè, S. V., Romeo, R., Romeo, G., Cappello, S., Bonsignore, M., and Denaro, R. (2015) 1,2,3-Triazole/MWCNT conjugates as filler for gelcoat nanocomposites: new active antibiofouling coatings for marine application. *Materials Research Express*, **2**(11), 3.
65. Finlay, J. A., Schultz, M. P., Cone, G., Callow, M. E., and Callow, J. A. (2013) A novel biofilm channel for evaluating the adhesion of diatoms to non-biocidal coatings. *Biofouling*, **29**(4), 401-411.
66. Schultz, M. P., Walker, J. M., Steppe, C. N., and Flack, K. A. (2015) Impact of diatomaceous biofilms on the frictional drag of fouling-release coatings. *Biofouling*, **31**, 759-773.
67. Watson, M. G., Scardino, A. J., Zalizniak, L., and Shimeta, J. (2015) Colonization and succession of marine biofilm-dwelling ciliate assemblages on biocidal antifouling and fouling-release coatings in temperate Australia. *Biofouling*, **31**, 709-720.
68. Muthukrishnan, T., Abed, R. M., Dobretsov, S., Kidd, B., and Finnie, A. A. (2014) Long-term micro fouling on commercial biocidal fouling control coatings. *Biofouling*, **30**(10), 1155-1164.
69. Kenawy, E. R., Worley, S. D., and Broughton, R. (2007) The Chemistry and applications of antimicrobial polymers: A state-of-the-art review. *Biomacromolecules*, **8**(5), 1359-1384.
70. *Textile Finishing*, Heywood, D. (ed.), Society of Dyers and Colorists, UK (2003).
71. Chung, D. W. and Lim, J. C. (2009) Study on the effect of structure of polydimethylsiloxane grafted with polyethylene oxide on surface activities. *Colloids and Surfaces A: Physicochemical and Engineering Aspects*, **336**, 35-40.
72. Kim, D. W., Noh, S. T., Jo, B. W. (2006) Effect of salt and pH on surface active properties of comb rake-type polysiloxane surfactants. *Collo-*

ids and Surfaces A: Physicochemical and Engineering Aspects, **287**, 106-116.
73. Dastjerdi, R., Montazer, M., and Shahsavan, S. (2009) Incorporation of texturing and yarn surface modification using nano sized colloidal particles. *Chemical Biology - Fundamental Problems of Bionanotechnology.*
74. Hofmann, H., Fink-Petri A., and Salaklang J., (2009) Nanoparticles for Diagnostic and Therapeutic Applications: The Potential of Superparamagnetic Iron Oxide Nanoparticles (SPION). *Proceedings of the 6th International Conference on Biomedical Applications of Nanotechnology,* Germany.
75. Geppert M., Hohnholt M., Grunwald I., Baeumer M., Dringen R., (2009) Accumulation of iron oxide nanoparticles in cultured brain astrocytes. *Journal of Biomedical Nanotechnology,* **5**(3), 285-293.
76. De Santa Maria, L. C., Souza, J. D. C., Aguiar, M. R. M. P., Wang, S. H., Mazzei, J. L., Felzenszwalb, I., and Amico, S. C. (2008) Synthesis characterization, and bactericidal properties of composites based on crosslinked resins containing silver. *Journal of Applied Polymer Science,* **107**, 1879-1886.
77. Petkova, P., Francesko, A., Fernandes, M. M., Mendoza, E., Perelshtein, I., Gedanken, A., and Tzanov, T. (2014) Sonochemical coating of textiles with hybrid ZnO/chitosan antimicrobial nanoparticles. *ACS Applied Materials and Interfaces,* **6**(2), 1164-1172.
78. Pan, Y., Xiao, H., Cai, P., and Colpitts, M. (2016), Cellulose fibers modified with nano-sized antimicrobial polymer latex for pathogen deactivation. *Carbohydrate Polymers,* **135**, 94-100.
79. Nogueira F., Vaz, J., Mouro, C., Piskin, E., and Gouveia, I. (2014) Covalent modification of cellulosic-based textiles: A new strategy to obtain antimicrobial properties. *Biotechnology and Bioprocess Engineering,* **19**(3), 526-533.
80. Farouk, A., Moussa, S., Ulbricht, M., Schollmeyer, E., and Textor, T. (2013) ZnO-modified hybrid polymers as an antibacterial finish for textiles. *Textile Research Journal,* **84**(1), 40-51.
81. Soraya, G., and Mortazavi, S. M. (2015) Microwave curing for applying polymeric nanocapsules containing essential oils on cotton fabric to produce antimicrobial and fragrant textiles, *Cellulose,* **22**(6), 4065-4075.
82. Gomes, A. P., Mano, J. F., Queiroz, J. A., and Gouveia, I. C. (2013) Layer-by-layer deposition of antimicrobial polymers on cellulosic fibers: a new strategy to develop bioactive textiles, *Polymers for Advanced Technologies,* **24**, 1005-1010.
83. Lynch, A. S., and Robertson, G. T. (2008), Bacterial and fungal biofilm infections. *Annual Review of Medicine,* **59**, 415-428.
84. Busscher, H. J., van der Mei, H. C., Subbiahdoss, G., Jutte, P. C., van den Dungen, J. A. M., Zaat, S. A. J., Schultz, M. J., and Grainger, D. W.

(2012) Biomaterial-associated infection: locating the finish line in the race for the surface. *Science Translation Medicine*, **4**(153), 110.
85. *The World Health Report 2004: Changing History*, World Health Organization (2004). Online: https://www.who.int/whr/2004/en/ [accessed 21st June 2019].
86. Boucher, H. W., Talbot, G. H., Bradley, J. S., Edwards, J. E., Gilbert, D., Rice, L. B., Scheld, M., Spellberg, B., and J., Bartlett (2009) Bad bugs, no drugs: No ESKAPE! An update from the Infectious Diseases Society of America. *Clinical Infectious Diseases*, **48**, 1-12.
87. Martinez, L. R., and Casadevall, A., (2007), Cryptococcus neoformans biofilm formation depends on surface support and carbon source and reduces fungal cell susceptibility to heat, cold, and UV light. *Applied Environmental Microbiology*, **73**, 4592-4601.
88. Shingu-Vazquez, M., and Traven, A. (2011), Mitochondria and fungal pathogenesis: drug tolerance, virulence, and potential for antifungal therapy. *Eukaryotic Cell, American Society of Microbiology*, **10**, 1376-1383.
89. Finkel, J. S., and Mitchell, A. P. (2011) Mitchell Genetic control of Candida albicans biofilm development. *Nature Reviews Microbiology*, **9**, 109-118.
90. Rodriguez-Emmenegger, C., Brynda, E., Riedel, T., Houska, M., Šubr, V., Alles, A. B., Hasan, E., Gautrot, J. E., and Huck, W. T. S. (2011) Polymer brushes showing non-fouling in blood plasma challenge the currently accepted design of protein resistant surfaces. *Macromolecular Rapid Communication*, **32**, 952-957.
91. *The Direct Medical Costs of Healthcare-Associated Infections in U.S. Hospitals and the Benefits of Prevention* (2009). Online: https://stacks.cdc.gov/view/cdc/11550 [accessed 21st June 2019].
92. Singh, B. (2012) Human pathogens utilize host extracellular matrix proteins laminin and collagen for adhesion and invasion of the host. *FEMS Microbiology Reviews*, **36**, 1122-1180.
93. Taresco, V., Crisante, F., Francolini, I., Martinelli, A., D'Ilario, L., Ricci-Vitiani, L., Buccarelli, M., Pietrelli, L., and Piozzi, A. (2015) Antimicrobial and antioxidant amphiphilic random copolymers to address medical device-centered infections. *Acta Biomaterialia*, **22**, 131-140.
94. Tran, P. A., and Webster, T. J. (2013) Antimicrobial selenium nanoparticle coatings on polymeric medical devices. *Nanotechnology*, **24**, 155101.
95. Muszanska, A. K., Rochford, E. T. J., Gruszka, A., Gruszka, A., Andreas, A. B., Henk, J. B., Willem, N., Henny, C. V., and Andreas, H. (2014) Antiadhesive polymer brush coating functionalized with antimicrobial and RGD peptides to reduce biofilm formation and enhance tissue integration. *Biomacromolecules*, **15**, 2019-2026.
96. Tyagi, M., and Singh, H. J. (2000) Iodinated P (MMA-NVP): An effi-

cient matrix for disinfection of water. *Journal of Applied Polymer Science*, **76**, 1109-1116.

97. Tan, S., Li, G., Shen, J., Liu, Y., and Zong, M. (2000) Study of modified polypropylene nonwoven cloth. II. Antibacterial activity of modified polypropylene nonwoven cloths. *Journal of Applied Polymer Science*, **77**(9), 1869-1876.

98. Sun, Y., and Sun, G. (2001), Durable and refreshable polymeric N-halamine biocides containing 3-(4'-vinylbenzyl)-5, 5-dimethylhydantoin. *Journal of Polymer Science, Part A: Polymer Chemistry*, **39**, 3348-3355.

99. Chen, Y., Worley, S. D., Kim, J., Wei, C.-I., Chen, T.-Y., Santiago, J. I., Williams, J. F., and Sun, G. (2003) Biocidal Poly(styrenehydantoin) beads for disinfection of water. *Industrial & Engineering Chemistry Research*, **42**, 280-284.

100. Duncan, R., and Kopeček, J. (1984) Soluble synthetic polymers as potential drug carriers. *Advances in Polymer Science*, **57**, 51-101.

101. Sharma, M., Padmavathy, N., Remanan, S., Madras, G., and Bose, S. (2016) Facile one-pot scalable strategy to engineer biocidal silver nanocluster assembly on thiolated PVDF membranes for water purification. *RSC Advances*, **6**, 38972-38983.

102. Ahmed, F., Santos, C. M., Mangadlao, J., Advincula, R., and Rodrigues, D. F. (2013) Antimicrobial PVK: SWNT nanocomposite coated membrane for water purification: Performance and toxicity testing. *Water Research*, **47**(12), 3966-3975.

103. Bollmann, U. E., Ou, Y., Mayer, P., Trapp, S., and Bester, K. (2015) Polyacrylate–water partitioning of biocidal compounds: Enhancing the understanding of biocide partitioning between render and water. *Chemosphere*, **119**, 1021-1026.

104. Zhang, L., Xu, J., Tang, Y., Hou, J., Yu, L., and Gao, C. (2016) A novel long-lasting antifouling membrane modified with bifunctional capsaicin-mimic moieties via in situ polymerization for efficient water purification. *Journal of Materials Chemistry A*, **4**, 10352-10362.

105. Sharma, G., Pathania, D., Naushad, M., and Kothiyal, N.C. (2014) Fabrication, characterization and antimicrobial activity of polyaniline Th(IV) tungstomolybdophosphate nanocomposite material: Efficient removal of toxic metal ions from water. *Chemical Engineering Journal*, **251**, 413-421.

106. He, J., Wang, W., Sun, F., Shi, W., Qi, D., Wang, K., Shi, R., Cui, F., Wang, C., and Chen, X. (2015) Highly efficient phosphate scavenger based on well-dispersed La(OH)$_3$ nanorods in polyacrylonitrile nanofibers for nutrient-starvation antibacteria. *ACS Nano*, **9**(9), 9292-9302.

107. Bastarrachea, L., Dhawan, S., and Sablani, S. S. (2011) Engineering properties of polymeric-based antimicrobial films for food packaging: A review. *Food Engineering Reviews*, **3**, 79.

108. Gemili, S., Yemenicioglu, A., and Altinkaya, S. A. (2009) Development

of cellulose acetate based antimicrobial food packaging materials for controlled release of lysozyme. *Journal of Food Engineering*, **90**(4), 453-462.
109. Dutta, P. K., Tripathi, S., Mehrotra, G. K., and Dutta, J. (2009) Perspectives for chitosan based antimicrobial films in food applications. *Food Chemistry*, **114**(4), 1173-1182.
110. Appendini, P., and Hotchkiss, J. H. (1997) Immobilization of lysozyme on food contact polymers as potential antimicrobial films. *Packaging Technology and Science*, **10**, 271-279.
111. Jin, T., and Zhang, H. (2008) Biodegradable polylactic acid polymer with nisin for use in antimicrobial food packaging. *Journal of Food Science*, **73**, M127-M134.

Index

A

abrasion resistance, 18, 21, 48, 50
active-passive approach, 110, 219
adhesion loss, 11
admittance, 206, 208
adsorption, 22, 35, 75, 77, 108, 115, 191, 193, 213-214
agglomeration, 76, 85, 117-118, 131-132
anodic passivation, 137
anti-fouling, 274, 278
aspect ratio, 110, 219, 236, 240

B

barrier properties, 14, 134, 138, 146, 149, 166, 188, 204, 207, 212, 219, 227, 229, 234, 241, 245, 261, 263, 265
biocompatibility, 45, 275, 282
blocking effect, 19, 146
bridging, 274

C

carbon nanotubes, 42, 73, 95, 110, 242, 266, 277
chain length, 62

Index

charge transfer resistance, 121, 123, 128, 131, 149, 190-191, 193, 202, 232

chitosan, 4, 30, 158, 165, 167, 215-216, 272, 275, 279, 284, 286, 289

coating capacitance, 201, 203, 231

coating resistance, 121-122, 128, 131, 149, 161

conducting polymers, 137, 164

constant phase elements, 121, 149, 232

contact angle, 44, 46, 171

corrosion protection efficiency, 210-211, 213

crosslinking, 3, 18, 21, 170, 172, 177, 179-180

crystal structure, 84-85

cytotoxicity, 272

D

degree of doping, 212

delamination, 9, 22, 34, 113, 171, 224, 227, 239, 248, 263

density, 5, 21, 31, 58, 69, 81, 87, 89, 103, 106, 141, 154, 170-171, 210, 237, 274, 277

dip coating, 54, 80, 94, 97, 113, 140, 222, 224, 244, 247-248

dopant, 145-146, 153, 163

double layer capacitance, 190, 192, 262

E

EDX, 85, 100, 223, 231-232

electrical resistivity, 23, 29

emeraldine base, 212

encapsulation efficiency, 111

equivalent circuit, 94, 121, 149, 160, 262-263

F

ferrites, 189, 215

flexibility, 7, 10, 16, 20, 225

fouling-release, 23, 35, 274, 285

free volume, 178, 254

FTIR, 8, 21-22, 78, 82, 97-98, 113, 117-118, 171, 176, 179,

Index

223, 237-238, 248, 251-253

G

glass transition temperature, 7, 11, 38, 170-172, 248, 264

H

half-cell reactions, 213
hardness, 17, 21, 26, 57-58, 70, 73, 171
heating cycle, 240, 250
humidity, 172, 176, 178, 184
hybridization, 73, 104, 236
hydrogels, 284
hydrogen bonding, 46, 65

I

ICPs, 137-138
immobilization, 64, 281, 289
in-situ polymerization, 208, 212
inductor, 151
inhibitor, 40, 49, 63, 67, 109-112, 114-116, 119, 121, 123-125, 128, 130-134, 138, 214, 241
ionic species, 138, 220, 233

L

layer by layer, 30, 114, 130, 215, 276
leaching, 43, 109, 282
low frequency impedance, 119-121, 128, 146, 158-159, 260, 264

Index

M

mechanical properties, 5-7, 9-11, 20-21, 27-29, 54, 143, 179-180, 188, 267-268
metal ions, 126, 132, 145, 231, 288
microbial strains, 279
microcapsules, 110, 133, 135
montmorillonite, 154, 167, 240-241, 264
morphology, 33, 54, 63, 77, 102, 142, 191, 193, 210-212, 225
multilayers, 276
multipotentiostat, 224

N

nano-casting, 18
nano-containers, 79, 110-112
network structure, 177, 227

O

oil and gas, 1-2, 8, 10, 19-20, 24, 26, 28-30, 34
optical properties, 171
oxide layer, 138, 146, 212, 247

P

pendant groups, 61
penetration, 94, 110, 153, 219
periodicity, 170
PLA, 165, 279
plasticization, 172
polarization resistance, 262
polyelectrolyte shells, 131
polymer-polymer contact, 188
pores, 74, 77-78, 93-94, 109, 122, 149, 153, 159, 178, 199, 202
porosity, 43, 122, 137, 170, 179, 199, 220
potentiodynamic polarization, 19, 114, 224
PPy, 137-138, 143, 145-146,

Index

151-155, 158, 160, 162-163, 167
precursor, 74, 82, 87, 100, 102
processability, 174

R

Raman spectra, 102, 141, 157, 163
relaxation, 193
reproducibility, 114, 141, 224, 249
resistor, 123, 149, 190, 196

S

salt spray, 23
scanning electron microscopy, 77-78, 171, 191, 275
scanning Kelvin probe, 22
Schottky barrier, 110, 213, 219, 236
segmental mobility, 254
semiconductor, 92, 237

sol-gel coatings, 110, 116, 135-136, 167, 243
steel coupons, 111, 113, 140, 190-191, 209, 222, 247
storage modulus, 246
superhydrophobicity, 33
surface area, 74, 78, 87, 101, 103, 190, 192, 202, 210, 227, 237
surface cracks, 142
surface ennobling, 137
surface passivation, 219
surfactants, 34, 87, 285

T

thermal resistance, 28, 110, 219, 245
thermally reduced graphene, 135, 214-215
thermogravimetric analysis, 223, 248
thermoset, 16, 18, 176, 182
titanium dioxide, 174
topography, 18, 211
transmission electron microscopy, 77-78, 198, 209, 223

U

UV absorber, 177
UV weathering, 113-114, 127-130
UV-cured, 182, 184

V

volatile organic compounds, 20

W

Warburg impedance, 232
water permeation, 11
wettability, 44

Y

Young's modulus, 143

CPSIA information can be obtained
at www.ICGtesting.com
Printed in the USA
LVHW051655240621
691069LV00002B/75